[illegible] v. [illegible] P. Eq [illegible] [illegible] up

1. [illegible] of [illegible] / [illegible] ?

2. [illegible] 2 [illegible] 3 → [illegible]

3. [illegible] ?

[illegible]

x 4

24m.

(v.)

x 1.30

[illegible] SALES

4. 2 [illegible] → [illegible] / FAST [illegible] - 70

v. [illegible] (+) [illegible] MODEL.

ACKNOWLEDGMENTS

The SFI Press would not exist without the support of William H. Miller and the Miller Omega Program.

Additional funding has been provided by Alana Levinson-Labrosse.

[COMPLEXITY ECONOMICS]

DIALOGUES OF THE APPLIED COMPLEXITY NETWORK I

Proceedings of
the Santa Fe Institute's
2019 Fall Symposium

W. BRIAN ARTHUR
ERIC D. BEINHOCKER
ALLISON STANGER

editors

© 2020 Santa Fe Institute
All rights reserved.

THE SANTA FE INSTITUTE PRESS

1399 Hyde Park Road
Santa Fe, New Mexico 87501

Complexity Economics:
Proceedings of the Santa Fe Institute's
2019 Fall Symposium
Dialogues of the Applied Complexity Network I

ISBN (PAPERBACK): 978-1-947864-35-1
Library of Congress Control Number: 2020949113

The SFI Press is supported by the
Miller Omega Program, with additional funding
provided by Alana Levinson-LaBrosse.

Buy low and sell high
Get rich and give to SFI
Complexity science saves the world.

—BILL MILLER

(A tongue-in-cheek, convention-busting haiku written during the Santa Fe Institute's 2019 Fall Symposium)

Orientation

Day 1: talks

OPPOSITE *A mosaic image of the American dollar bill*

Day 2: Panels

Appendix

ORIENTATION

According to economist Stephen Ziliak, **haiku economics** originated in 2001, when he introduced the subject to his students at the Georgia Institute of Technology. The idea was to put the principles of economics together with the principles of haiku for the mutual study and benefit of both arts and sciences.

Drawing on this concept—which continues to be used by poetic economists around the globe—the organizers of SFI's New Complexity Economics Symposium invited participants to submit haiku during the event. Examples of these are presented throughout this volume.

Human flourishing

deadly externalities

THE CUSP OF WINTER

—TIM HODGSON

FOREWORD: INTERACTION & INNOVATION

William Tracy & Casey Cox

Geoffrey West, SFI faculty member and past president, observed in his book *Scale* that the "economic output, the buzz, the creativity, and culture of a city or company all result from the nonlinear nature of the multiple feedback mechanisms embodied in the interactions between its inhabitants, their infrastructure, and the environment."[1] In other words, unscripted interactions between diverse inhabitants seem particularly important for driving innovation and growth in cities and other human organizations. These insights are among the organizing principles behind SFI's Annual Applied Complexity Network (ACtioN) and Board of Trustees Symposium, on which this book is based, and indeed the entire *Dialogues of the Applied Complexity Network* series, as well.

Each November, ACtioN gathers a diverse group of scholars, practitioners, leaders, and artists with interests in a particular theme, and then creates an environment designed to facilitate rare, serendipitous but valuable interactions that birth new ideas and insights. Formal talks and panels play a role in stimulating smaller discussions and easing the social constraints that often frustrate productive conversations. Unlike ordinary academic conferences, we invite talks from leading scholars as well as from artists, practitioners, and other mavericks who will drive our thinking in novel directions. Symposium talks reveal insights from the frontiers of knowledge, but do not collectively represent the totality of a topic or field. Like SFI's monastic setting in the mountains above

1 West 2017, p. 22

OPPOSITE *Economic Curiosity: Ambergris*

Santa Fe, the symposia serve to inspire a collaborative exploration of a timely and important subject from new perspectives.

These symposia reflect ACtioN's broader mission. ACtioN is a community through which SFI collaborates with practitioners at firms, governments, and nonprofits that seek to deploy complex-systems insights to achieve their goals. By examining a variety of complex systems from different perspectives, ACtioN members gain new understandings and SFI researchers gain new insights. In addition to the annual symposium, ACtioN organizes a host of online and in-person topical meetings, in which SFI researchers and ACtioN practitioners discuss how to apply emerging ideas from complexity science. ACtioN also offers members opportunities to participate in SFI education programs, with the aim of promoting not only an understanding of but also organizational fluency with complexity science.

The *Dialogues of the Applied Complexity Network* book series supports ACtioN's efforts to facilitate ongoing conversations. *Complexity Economics*, the first volume in the series, is a record of the talks, panels, and full-group discussions at the Fall 2019 symposium. These conversations, as well as the reflections written by select participants after the meeting, highlight ways in which SFI's practitioner community is likely to apply complexity science in the near future. This volume also reveals the symposium's lighter side. For example, the images that appear at the beginning of each chapter in this volume come from economics-themed mosaic artwork that decorated each table at the event, and the haiku were written by attendees during the symposium. These proceedings will be useful for participants reflecting on the conversations in which they took part, individuals looking to join SFI's broader practitioner community, and those seeking to understand the historic trajectories of our conversations.

In November 2019, more than 150 ACtioN members, SFI researchers, and others from the SFI community gathered for the 26th annual Applied Complexity Network Symposium. The two-day conference focused on *New Complexity Economics*, a nod to an earlier symposium addressing the same topic. By moving away from the assumption of universal rationality and incorporating feedback loops, complexity economics offers a dynamic representation of economic systems. This stands in contrast to the more equilibrium-oriented view of traditional economics.

Complexity economics constitutes one of SFI's major contributions to the world. Yet, even fifteen years ago, the core tenets of complexity economics were viewed with skepticism by the mainstream economics community. A litany of major policy failures has since sforced large swaths of the economics community to acknowledge the validity of critiques leveled by complexity economics. The 2007–2008 finance crisis arguably served as a tipping point in the loosening grip of the twentieth century's reductionist economic orthodoxy. However, there were also earlier rumblings, notably the failure of Eastern Europe's economic transition, whose design was governed by the principles of orthodox economics. This failure stood in particularly sharp contrast to the success of China's economic transitions, whose logic was largely congruent with complexity economics.

Unlike mainstream academia, practitioners from government, industry, and civil society have long been comfortable with complexity economics. This is perhaps most evident at technology firms, where complexity economic thinking was quickly proven to be a useful framework, particularly in regards to increasing returns to scale and innovation. There has been a recent groundswell of popular interest in these ideas, as analysts rush to provide narratives for the technology sector's increasing dominance within the economy. As complexity economics achieves a new level of prominence within both academia and practice, this is a poignant moment to review its most salient contributions and promising directions for future growth.

The rest of this volume—in particular, the Editors' Introduction and Chapter 7, "Complexity Economics: Why Does Economics Need This Different Approach," by W. Brian Arthur—provides a deeper discussion of the history and future of complexity economics.

COMPLEXITY ECONOMICS: AN INTRODUCTION

W. Brian Arthur, Eric D. Beinhocker & Allison Stanger

In September 1987, the then-nascent Santa Fe Institute (SFI) held a conference to look at the economy as an evolving complex system. It was organized by physicists Philip Anderson and David Pines and economist Kenneth Arrow.[1] Out of that conference a year later came the Santa Fe Institute's first research program: The Economy as an Evolving Complex System. One of us (Arthur) headed that program, and another (Beinhocker) joined it in 1994, and from that program in turn came a different approach to economics—what many refer to as "complexity economics."[2] SFI has been associated with complexity economics ever since, and the approach has continued to develop and influence not just economics, but social science more broadly. In 2014 SFI devoted its annual symposium to complexity economics, and it did so again in 2019. The collection of talks in this volume provides a snapshot of recent thinking and examples of how the initial ideas have developed over the decades.

THE EMERGENCE OF COMPLEXITY ECONOMICS

Complexity economics views the economy as a system not necessarily in equilibrium, but rather as one where agents constantly change their actions and strategies in response to the outcomes they mutually create. It holds that computation as well as mathematics is useful

1 See Ole Peter's essay, "Remembering Three SFI Thinkers," on page 213 for a vignette of Kenneth Arrow.

2 The term "complexity economics" was first introduced by Arthur (1999).

OPPOSITE *Economic Curiosity: Red Paper Clip*

in economics, that increasing as well as diminishing returns may be present in an economic situation, and that the economy is not something given and existing, but forms from a constantly developing set of actions, arrangements, and technological innovations. The economy is thus comprised of evolving networks of interacting agents, institutions, and technologies—networks of networks. The macro- level patterns of the economy—growth, innovation, business cycles, market booms and busts, inequality, and carbon emissions—then emerge from these dynamic micro- and meso-level interactions. From the complexity economics perspective, change is largely an endogenous phenomenon, not simply the result of unexplained shocks from outside the system.

[Adam] Smith's famous metaphor of the "invisible hand" . . . was a statement about emergence—how individual actions "without intending it, without knowing it" lead to collective outcomes, which feed back to influence further actions.

The complexity viewpoint has modern roots in pioneering work in the physical sciences conducted decades ago by groups in Brussels, Stuttgart, Ann Arbor, Los Alamos, and elsewhere. But the ideas have even earlier precedents in economics. Adam Smith had a deep, intuitive understanding of emergence and was arguably the first complexity economist. Smith and other early economists were aware that aggregate patterns emerge from individual behavior and interactions, and that individual behavior responds to these aggregate patterns. Smith's famous metaphor of the "invisible hand" of markets is popularly misinterpreted as a message that "greed is good" (something Smith did not believe), but in reality was a statement about emergence—how individual actions "without intending it, without knowing it" lead to collective outcomes which feedback to influence further actions.[3]

3 For a recent discussion see Norman (2018), pp. 154–159.

There is thus a recursive, reflexive loop at the heart of the economy.[4] Complexity economics asks how this loop drives the behavior of the system over time, i.e., how will the pattern of the system today shape individual decisions which will then collectively create the pattern of the system tomorrow.

This is an obvious question, but a difficult one, and in the nineteenth century economists chose to ask a simpler question that was more tractable with the tools available at the time. They asked instead what individual behaviors (actions, strategies, expectations) would be upheld by—would be *consistent with*—the aggregate patterns they create. What patterns would call for no changes in micro-behavior, and so would be in equilibrium?[5] This equilibrium strategy was a natural way to analyze problems in the economy and render them open to mathematical analysis. But it placed a strong filter on what one could see. Under equilibrium by definition there isn't scope for exploration, or forming new structures, or creation; so, anything in the economy that takes adjustment—innovation, structural change, history itself must get dropped from the theory. The resulting analysis may be elegant, but ever since its nineteenth-century origins, heterodox economists ranging from Joseph Schumpeter to Friedrich Hayek, Herbert Simon, Kenneth Boulding, and Elinor Ostrom have highlighted its limitations.

It was within this context that SFI's economics program started in 1988. After much discussion, the group at SFI decided to ask: what would economics look like if we allowed nonequilibrium? How do individual actors in the economy react, make decisions, anticipate, and strategize, in response to the pattern they have created? And what kinds

4 See the 2013 *Journal of Economic Methodology*, 20(4), symposium on reflexivity and the economy, notably George Soros, "Fallibility, reflexivity, and the human uncertainty principle," pp. 309–329.

5 General equilibrium theory asks what prices and quantities of goods produced and consumed would be consistent with—would pose no incentives for change to—the overall pattern of prices and quantities in the economy's markets. Classical game theory asks what strategies, moves, or allocations would be consistent with—would be the best course of action for an agent (under some criterion)—given the strategies, moves, allocations his rivals might choose. And rational expectations economics asks what expectations would be consistent with—would on average be validated by—the outcomes these expectations together created.

of networks, institutions, technologies, artifacts, and patterns emerge and evolve?

How to answer these questions and what methods to use weren't obvious at first. Standard (neoclassical) economics assumes identical agents solving an identical problem using rational logic, knowing that other agents are identical and face the same problem. But in the actual economy this is rarely true. Different entrepreneurs starting up tech companies might not know how well their individual technologies will work, how the government will regulate them, or who their competitors will be. They would be subject to fundamental uncertainty, and so the problem they face is not well-defined. It follows that rationality is not well-defined either; there can't be a logical solution to a problem that isn't logically defined. There is no "optimal" move.

People of course do act in ill-defined situations all the time. They form individual hypotheses—internal models—about the situation they're in and continually update them. They constantly adapt or discard their hypotheses, strategies, and the actions as they explore.[6] They proceed in other words by induction.[7] As it happened, John Holland, a superb computer scientist and cognitive theorist, was part of SFI's early program and had developed methods by which computer programs could "learn" in games by creating hypotheses and adapting or discarding or mutating them. We could use this idea, we realized, as the basis for a general method to investigate situations in economics that weren't well-defined and weren't in equilibrium.

Several things came out of this approach. In problem after problem we saw a world where beliefs, strategies, and actions of agents were being "tested" for performance and survival within an "ecology" that these beliefs, strategies, and actions together created. The behaviors themselves were adapting to the ecology they together created and that ecology in turn evolved as the behaviors changed. A distinct biological evolutionary theme emerged (another area in which the SFI community had deep expertise). While economic systems have important differences from biological ones, they share some deep commonalities—notably, both use energy to create islands of order, stability, and

6 See for example Hommes (2011).

7 See Holland et al. (1986); Sargent (1993); Arthur (1994).

local equilibrium (or homeostasis) in larger seas of disorder, but those islands of local stability are by necessity evolving as their environment evolves, meaning that the system as a whole is highly dynamic and far from equilibrium.[8] Modeling such a world requires tools that are different from the standard economic approaches and our team at SFI began to put together computational methods to look at economic problems.[9] These models were built "bottom-up" around agents' individual behaviors, and so they used an early form of agent-based modeling. This is now a well-known method that has been applied to a variety of economic problems.[10]

People do of course act in ill-defined situations all the time. They form individual hypotheses—internal models—about the situation they're in and continually update them. They constantly adapt or discard their hypotheses, strategies, and the actions as they explore. They proceed, in other words, by induction.

But computers aren't the only way to understand complex adaptive systems, and the desire to understand the messy reality of human social systems drove innovation in other methods as well. For example, a community of researchers centered at SFI has, over decades, integrated results from laboratory behavioral experiments, field experiments,

8 Beinhocker (2006).

9 This early work on computational economic models at SFI was led by economists, including W. Brian Arthur, Blake LeBaron, with help from computer scientist John Holland, statistician David Lane, and physicist Richard Palmer. Early on the approach was called "element-based modeling" and then later the more descriptive "agent-based modeling" came into use.

10 For other early examples of agent-based modeling in economics see Epstein and Axtell (1996); Amman et. al. (1996); and Tesfatsion and Judd (2006).

cross-species studies, archaeological and anthropological data, models of cultural evolution, and innovations in game theory to generate deep insights into the role of human prosocial behaviors and cooperative norms and institutions in the creation and evolution of economies.[11] Such multi-method, interdisciplinary work, while a core part of SFI's *raison d'être*, was not typical of economics as a field at the time and still remains relatively rare.

FROM CRISES TO OPPORTUNITY

Max Planck famously remarked that science advances one funeral at a time, but as Thomas Kuhn observed, it also advances one crisis at a time. While the rational-actor, efficient-market, equilibrium view of economics traces its roots to the nineteenth-century work of figures such as Léon Walras, its modern ascendancy in academia and policymaking has its origins in the economic crises of the 1970s. The collapse of the postwar Bretton Woods regime, a series of currency crises, high inflation, recessions, and high unemployment in many Western countries caused policymakers to lose faith in Keynesian analyses and prescriptions. This created an opening for the ideas of monetarists, led by Milton Friedman, and neoclassical economists such as Robert Lucas, Thomas Sargent, Eugene Fama, and Gary Becker. Their ideas had a profound influence from the 1970s to 2000s, shaping not just academic research, but also what many historians refer to as the neoliberal policy consensus of the period. That intellectual consensus in turn drove agendas on deregulation, market liberalization, free trade, financialization, increased reliance on economic incentives to shape behavior, and attempts to scale back the role of the state.[12]

While this academic and policy consensus held for several decades, the first crisis of the neoliberal era appeared early on and provided the motivations for the founding of the SFI economics program. Following the neoliberal recipe book, many Latin American countries opened up their economies to international capital flows, but by 1982 found themselves gripped by a cascade of currency crises and debt defaults. One of

11 See for example Gintis, Bowles, Boyd, and Fehr (2005); Bowles and Gintis (2011); Bowles (2016); and Gintis (2017).

12 See Burgin (2012).

the banks with the greatest exposure was Citicorp, and its then-chairman and CEO, John Reed, began a quest to understand what had caused the crisis and how future crises could be prevented. Reed consulted widely with leading economists, but the rational-actor, efficient-market theories espoused at the time offered little insight into the formation and bursting of such a massive bubble—debt crises are a profoundly disequilibrium phenomenon. Reed's quest eventually brought him to SFI, and Citicorp provided funding for the 1987 founding workshop and the 1988 start-up of the SFI economics program. One can thus say that complexity economics was born out of crisis.

Max Planck famously remarked that science advances one funeral at a time, but as Thomas Kuhn observed, it also advances one crisis at a time.

With hindsight, the 1980s Latin American and 1990s Asian debt crises were just warm-ups for the big crisis to come. The 2008 global financial crisis effectively ended the period of neoclassical/neoliberal intellectual dominance. In an interview after the crash, US Treasury Secretary Hank Paulson was asked whether economics had been of any help during the crisis and bluntly replied, "The economic models were worthless."[13] Likewise, Jean-Claude Trichet, president of the European Central Bank at the time, said, "we felt abandoned by the conventional tools," and called for "inspiration from other disciplines: physics, engineering, psychology, biology," where "scientists have developed sophisticated tools for analyzing complex dynamic systems in a rigorous way."[14]

Such calls from policymakers gave a new impetus and urgency to the complexity-economics research program. Significant effort by multiple research groups was invested in applying complex-systems techniques to

13 Tapper (2010).

14 Trichet (2010).

better understanding issues of financial system stability and contagion.[15] This work opened doors to new collaborations with policymakers, and complexity economics ideas were actively debated, and experimented with inside institutions that had once been bedrocks of neoclassical thinking, including major central banks as well as international institutions such as the International Monetary Fund and World Bank.[16] The Organisation for Economic Cooperation and Development launched a major effort called "New Approaches to Economic Challenges" (NAEC) with the explicit mission of bringing complexity-economics thinking into policymaking.[17] One of the biggest casualties of the 2008 crisis was confidence in the equilibrium, micro-founded, macroeconomic models often used by policymakers (so-called dynamic stochastic general equilibrium, or DSGE, models).[18] A number of initiatives were launched to explore alternatives, including those offered by complexity economics.[19]

Other policy challenges have provided motivations for advances in complexity economics. The remarkable pace of technological change and its impacts on society have prompted fundamental explorations into the nature of technology, innovation processes, and the co-evolution of technologies and institutions.[20] A complex-systems perspective has also been increasingly applied to questions of economic growth,

---

15 For example, see the EU funded Project CRISIS led by Domenico Delli Gatti (Project Director), J. Doyne Farmer (Scientific Director), Jean-Philippe Bouchaud, Mauro Gallegati, Cars Hommes, Giulia Iori, Fabrizio Lillo, Stefan Thurner, and Eric Beinhocker. See the *Journal of Economic Dynamics & Control*, 50, 2015, special issue on "Crises and Complexity" for a collection of papers.

16 For examples of the application of complexity economics to Bank of England policymaking see Baptista, Farmer, Hinterschweiger, Low, Tang, and Uluc (2016) and Farmer, Kleinnijenhuis, Nahai-Williamson, and Wetzer (2020).

17 See www.oecd.org/naec. SFI external faculty member Alan Kirman has played a key role as an advisor to the OECD NAEC initiative.

18 See for example *Oxford Review of Economic Policy*, vol. 34, issue 1–2, spring-summer 2018, special issue on Rebuilding Macroeconomic Theory.

19 See for example the UK Economic and Social Research Council (ESRC) Rebuilding Macroeconomics initiative, rebuildingmacroeconomics.ac.uk.

20 See Arthur (2009). In 2017 and 2018 SFI hosted workshops on the co-evolution of physical and social technologies, see Farmer, Markopoulou, Beinhocker, and Rasmussen (2020) for a summary and eudemonicproject.org for more information.

development, and inequality.[21] And finally, it is becoming increasingly clear that the existential question facing humankind—how to create an economy that delivers human well-being without causing planet-wide ecological collapse—can best be understood through an interdisciplinary complexity-economics perspective.[22]

Science is always a work in process, and complexity economics still has much maturing to do, but over the thirty-three years that separates the first SFI economics workshop and the symposium summarized in this volume, much progress has been made. The complexity-economics community has grown well beyond its origins at SFI, with individual scholars pursuing this approach around the world, and highly active research groups in the US, UK, Italy, France, the Netherlands, Switzerland, Austria, Singapore, and other countries. There have been numerous advances in theory, modeling, data, and methodologies, drawn from across a large range of disciplines. While complexity economics is still a new perspective in economics, many of its key ideas have entered the mainstream of the field. Most economists now accept the need for more realistic assumptions in economic theory; and methodologies such as network analysis, heterogeneous agent models, evolutionary game theory, and economic lab experiments are becoming standard. Understandably, there still remains a reluctance to let go of the equilibrium framework, embrace ideas of evolution and emergence, and make full use of modern data and computational tools. But in our view, if economics is to be relevant to the challenges we face—from pandemics to economic inequality, technology acceleration, and climate change—it will have to embrace the complexity of the phenomena it seeks to understand.

THE SYMPOSIUM

The November 2019 conversations in Santa Fe that follow were not intended to be a comprehensive review of the current state of complexity economics. Rather they cover a diverse array of topics, viewpoints, and ideas on future directions for research. The participants include

21 See for example Hidalgo and Hausmann (2009); Hausmann and Hidalgo (2011); and Banerjee and Yakovenko (2010).

22 See for example Farmer et al. (2015, 2016, 2019); Brand-Correa and Steinberger (2017); and O'Neill et al. (2018).

members of the SFI academic community as well as SFI's Applied Complexity Network (ACtioN), which brings together the worlds of business, finance, technology, and government. Thus, what follows in the volume are edited transcripts of the talks and panels rather than more traditional academic papers. We hope that this format makes them accessible to a wide audience.

. . . if economics is to be relevant to the challenges we face—from pandemics, to economic inequality, technology acceleration, and climate change—it will have to embrace the complexity of the phenomena it seeks to understand.

In his introductory remarks (Chapter 1), SFI President David Krakauer sets the stage tracking developments in the core themes in economics from the Enlightenment into the twenty-first century, from the mechanics of Newton to concepts of agency, adaptation, and purpose brought in by Adam Smith and Charles Darwin, to the impacts of economics on the real world, exemplified by the work of John Maynard Keynes. As Krakauer notes, the concern of economics is now "the global stability of the planet" and the stakes are high.

Eric Beinhocker, picking up on Krakauer's historical theme, traces the development of the discipline of economics and how its core theoretical insights regarding the imagined reality of a self-regulating market economy were translated into the ideology of neoliberalism, which has become destructive for both humans and nature. Neoliberal ideology has supported a "greed is good" popular and political culture based on an overly simplistic, pseudoscientific view that self-interested action would, through the magic of markets, automatically produce societal betterment and human flourishing. Beinhocker argues in Chapter 2 that it is not atomistic competition in markets that has made societies prosperous, but rather human cooperation and prosocial behaviors in an evolving ecology of market and non-market institutions. True prosperity comes from "solving human problems", and solving complex problems requires sustained cooperation at scale. The state, business, and civil society all have roles to play in creating the inclusion, fairness, and trust necessary for sustaining cooperation at scale. Furthermore, Beinhocker argues, the environment is not an "externality" but deeply intertwined with human social systems, and solving the problem of creating a zero-carbon, sustainable economy is not a "cost" to be traded off with future "benefits" but an existential necessity that also is a historic opportunity to increase human and non-human well-being. Beinhocker and his collaborator Nick Hanauer see the potential for complexity economics to underpin a new economic ideology they call "market humanism," an ideology that reinforces rather than undermines the ties that bind humans together in the cooperative enterprise of civilization.[23]

In Chapter 3, Allison Stanger considers the impact of neoliberal understandings of the global political economy on the American loss of faith in the efficacy of government. When government is assumed to be an obstacle to economic growth, the privatization of what were previously inherently governmental functions follows. The hollowing out of government blurs the demarcation between the public and private sectors, undermining the capacity of government to serve all of its citizens. When elites benefit at the expense of the majority, the obsessive focus on efficiency blinds the powerful to the political ramifications of growing social inequality and mass loss of trust in government. Technological innovation has accelerated these trends, so that the very Enlightenment values that produced spectacular prosperity are now increasingly seen as rationalizations of domination. Whatever new national story Americans together construct must speak to the realities of both the winners and the losers of economic development and value the contributions of care for others as well as unbridled self-interest.

Ole Peters (Chapter 4) introduces the non-initiated to "ergodicity economics," which postulates that individuals do not optimize their expected utility value over a set of well-defined possible probabilistic outcomes, but instead consider their decisions over time in the face of uncertainty. This has profound and underappreciated implications

23 Beinhocker and Hanauer (forthcoming).

for modeling economic decision making and systems as a core assumption of standard economics is that the time average and the expected (ensemble) value of an observable are one and the same. In real-world situations they often are not, and human decision making has evolved to help us make decisions in a world where outcomes are path dependent and where we only get to live in one future (in contrast the standard expected utility model can be interpreted as one where history doesn't matter and where we can probabilistically live in multiple future universes). Peters has applied this thinking to a range of problems including optimal levels of debt, market efficiency, and inequality.

Matthew O. Jackson's work on markets and social networks, a collaboration with the Nobel Laureates Abhijit Banerjee and Esther Duflo, highlights the importance of being attentive to the ways in which economic and social systems interact (Chapter 5). Studying economic relationships in a manner that detaches them from lived experience in particular communities can produce misleading inferences about what development interventions are effective and why. Jackson partnered with BSS bank in India, who was interested in securing better participation in their microfinance, Grameen Bank–style loans given to women aged eighteen to fifty. The bank's strategy was to target well-connected individuals to maximize the likelihood of diffusion. Jackson considered different measures of this centrality that might be used to predict outcomes. He found that informal networks made a difference in spreading information about opportunities, but that once the loans were dispensed, and people were the recipients of microfinance, it changed the structure of their social network and interactions. This suggests that while human relationships can be exploited to expand markets, the spread of markets can also spawn unintended consequences for the social fabric, potentially increasing inequality rather than ameliorating it.

C. Mónica Capra (Chapter 6) considers the collective action problem of protests, where an individual does not benefit from a particular action unless others act in similar fashion. Social media would seem to transform the dynamics of this interaction, since a human with anger and grievances can easily find like-minded compatriots. Narratives can be generated through online relationships that may or may not be products of the fact-based world. Using Amazon Mechanical Turk, Capra's team generated two communications experiments, one in which a single human subject is communicating with four bots, and a second comprised of five humans. Three types of communication were studied in each experiment—no communication, wall communication, and bilateral communication, all under different local network conditions. Bilateral communication improved the decisions of individuals, in that it was most conducive to developing a common narrative that is a necessary condition for successful coordination.

In Chapter 7, Brian Arthur's closing talk provides a firsthand account of the founding of the Santa Fe Institute and of complexity economics from someone present at the creation of both. Traditional equilibrium economics assumes that stasis, rather than adaptation and change, is life's predominant condition, and that the economy is comprised of identical hyper-rational agents. Based on a set of assumptions that enable mathematization of the core problems of economic life, it facilitated elegant solutions, fine for presentation in a textbook but not always appropriate for characterizing the realities of economic life. Complexity economics lets go of these restrictive assumptions. It assumes differing agents who are forced to make decisions in the presence of not knowing what other agents will do—in the presence therefore of fundamental uncertainty. They thus explore and adapt constantly, and this renders non-equilibrium as the norm in the economy, not the exception. It is a bottom-up approach that incorporates both space and the reality of events triggering further events in the unfolding of the economy. Arthur encourages us to see the economy not as a well-oiled machine but instead as an ecology of mutually adapting beliefs, strategies, and actions.

The second day of the workshop, documented here in Chapters 8 through 12, featured lively and engaging interdisciplinary discussion of both complexity theory and its past and future deployment in grappling with challenging real-world financial and socioeconomic problems. The offerings commenced with discussion of computational approaches to complex economies and moved on to an exploration of the leverage that physics might provide in apprehending economic systems. SFI President David Krakauer moderated both panels. Biology's contributions to our understanding of the economic organism and the insights of the social sciences for appreciating the significance of economic architectures were next on the docket. SFI Vice President for Applied Complexity William Tracy guided these conversations. Wall Street Journal reporter Paul J. Davies oversaw the final panel, comprised of successful practitioners who considered the impact of complex adaptive systems thinking born at SFI on the evaluation of investment alternatives.

We thank David Krakauer for his able chairing of the symposium, Casey Cox and the SFI Press for editing the transcripts that follow, and the SFI staff for all of their support on the symposium and this volume.

Shortly after the symposium, the first cases of COVID-19 were detected in Wuhan, China, and the world soon plunged into a crisis unprecedented in modern times. The pandemic has vividly illustrated the need to understand the complex, dynamic, and highly interconnected nature of our economic, social, and bio-physical systems. Other major challenges, from the climate crisis, to fracturing democracies, accelerating technology change, and future issues that we aren't yet aware of, will require the insights that only a complex-systems perspective can bring. Complexity economics has made much progress over the past three decades, but the work has only just begun.

W. Brian Arthur
Eric D. Beinhocker
Allison Stanger

References

Amman, Hans M., David A. Kendrick, and John Rust, eds. 1996. *Handbook of Computational Economics*. Vol. 1. Amsterdam: Elsevier.

Arthur, W. Brian. 2009. *The Nature of Technology: What It Is and How It Evolves*. New York, NY: Free Press.

———. 1994. "Inductive Reasoning and Bounded Rationality." *The American Economic Review* 84 (2): 406–11.

———. 1999. "Complexity and the Economy." *Science* 284 (5411): 107–9.

Banerjee, Anand, and Victor M. Yakovenko. 2010. "Universal Patterns of Inequality." *New Journal of Physics* 12 (7): 075032.

Baptista, Rafa, J. Doyne Farmer, Marc Hinterschweiger, Katie Low, Daniel Tang, and Arzu Uluc. 2016. "Macroprudential Policy in an Agent-Based Model of the UK Housing Market." Bank of England Staff Working Paper, No. 619.

Beinhocker, Eric D. 2006. *The Origin of Wealth: Evolution, Complexity, and the Radical Remaking of Economics*. Boston, MA: Harvard Business School Press.

Beinhocker, Eric D., and Nick Hanauer. Forthcoming. *Market Humanism*.

Bowles, Samuel. 2016. *The Moral Economy: Why Good Incentives Are No Substitute for Good Citizens*. New Haven, CT: Yale University Press.

Bowles, Samuel, and Herbert Gintis. 2011. *A Cooperative Species: Human Reciprocity and Its Evolution*. Princeton, NJ: Princeton University Press.

Brand-Correa, Lina I. and Julia K. Steinberger. 2017. "A Framework for Decoupling Human Need Satisfaction from Energy Use." *Ecological Economics* 141: 43–52.

Burgin, Angus. 2012. *The Great Persuasion: Reinventing Free Markets since the Depression*. Cambridge, MA: Harvard University Press.

Epstein, Joshua M. 2009. "Modelling to Contain Pandemics." *Nature* 460 (August): 687.

Epstein, Joshua M., and Robert Axtell. 1996. *Growing Artificial Societies: Social Science from the Bottom Up*. Cambridge, MA: MIT Press.

Eudemonic Project. 2019. https://eudemonicproject.org.

Farmer, J. Doyne, Cameron Hepburn, Penny Mealy, and Alexander Teytelboym. 2015. "A Third Wave in the Economics of Climate Change." *Environmental Resource Economics* 62: 329–357.

Farmer, J. Doyne, Cameron Hepburn, Matt C. Ives, Thomas Hale, Thomas Wetzer, Penny Mealy, Ryan Rafaty, Sugandha Srivastav, Rupert Way. 2019. "Sensitive Intervention Points in the Post-Carbon Transition." *Science* 365 (6436): 132–134.

Farmer, J. Doyne, Alissa M. Kleinnijenhuis, Paul Nahai-Williamson, and Thom Wetzer. 2020. "Foundations of System-Wide Financial Stress Testing with Heterogeneous Institutions." Bank of England Working Paper No. 861.

Farmer, J. Doyne, and François Lafond. 2016. "How Predictable Is Technological Progress?" *Research Policy* 45 (3): 647–65.

Farmer, J. Doyne, Fotini Markopoulou, Eric D. Beinhocker, and Steen Rasmussen. 2020. "Collaborators in Creation." Aeon, February 11, 2020. https://aeon.co/essays/how-social-and-physical-technologies-collaborate-to-create

Gintis, Herbert. 2017. *Individuality and Entanglement: The Moral and Material Bases of Social Life*. Princeton, NJ: Princeton University Press.

Gintis, Herbert, Samuel Bowles, Robert Boyd, and Ernst Fehr, eds. 2005. *Moral Sentiments and Material Interests: The Foundations of Cooperation in Economic Life*. Cambridge, MA: MIT Press.

Hausmann, Ricardo, and César A. Hidalgo. 2011. "The Network Structure of Economic Output." *Journal of Economic Growth* 16 (4): 309–42.

Hidalgo, César A., and R. Hausmann. 2009. "The Building Blocks of Economic Complexity." *Proceedings of the National Academy of Sciences* 106 (26): 10570–75.

Holland, John, Keith J. Holyoak, Richard E. Nisbett, and Paul R. Thagard. 1986. *Induction*. Cambridge, MA: MIT Press.

Hommes, Cars. 2011. "The Heterogeneous Expectations Hypothesis: Some Evidence from the Lab." *Journal of Economic Dynamics and Control* 35 (1): 1–24.

Hommes, Cars, and Giulia Iori, eds. 2015. "Crises and Complexity." *Journal of Economic Dynamics & Control*, special issue, 50: 1–202.

Norman, Jesse. 2018. *Adam Smith: What He Thought and Why It Matters*. London, UK: Allen Lane.

OECD: New Approaches to Economic Challenges. 2020. https://www.oecd.org/naec.

O'Neill, Daniel W., Andrew L. Fanning, William F. Lamb, and Julia K. Steinberger. 2018. "A Good Life for All within Planetary Boundaries." *Nature Sustainability* 1 (2): 88–95. https://doi.org/10.1038/s41893-018-0021-4.

Pichler, Anton, Marco Pangallo, R Maria Rio-Chanona, François Lafond, and J. Doyne Farmer. 2020. "Production Networks and Epidemic Spreading: How to Restart the UK Economy?" *Covid Economics*, no. 28 (May): 79–151.

Rebuilding Macroeconomics. 2017. https://rebuildingmacroeconomics.ac.uk.

Vines, D. et al. 2018. "Rebuilding Macroeconomic Theory." Special Issue in *Oxford Review of Economic Policy* 34 (1–2).

Sargent, Thomas J. 1993. *Bounded Rationality in Macroeconomics*. Oxford, UK: Clarendon Press.

Soros, George. 2013. "Fallibility, Reflexivity, and the Human Uncertainty Principle." *Journal of Economic Methodology* 20 (4): 309–29.

Tapper, Jake. 2010. "After the Crash: Former Treasury Secretary Hank Paulson '68 Looks Back on His Role during the Worst Financial Crisis since the Great Depression." *Dartmouth Alumni Magazine*, April 2010. https://dartmouthalumnimagazine.com/articles/after-crash.

Tesfatsion, Leigh, and Kenneth L. Judd, eds. 2006. *Handbook of Computational Economics: Agent-Based Computational Economics.* Vol. 2. Amsterdam, Netherlands: Elsevier.

Trichet, Jean-Claude. 2010. "Reflections on the Nature of Monetary Policy Non-Standard Measures and Finance Theory." Speech to the ECB Annual Conference, November 18. https://www.ecb.europa.eu/press/key/date/2010/html/sp101118.en.html.

DAY 1: TALKS

After Arthur's talk

I'll visit EL FAROL BAR

unless it's busy

—ANONYMOUS ATTENDEE

The El Farol Bar problem, created in 1994 by W. Brian Arthur, is a problem in game theory in which a fixed population wants to go have fun at El Farol, unless it's too crowded.

THE
BUTCHER

9

COMPLEX ECONOMIES: FROM THE KEYNESIAN ORBIT TO THE DARWINIAN WORM

David C. Krakauer

> "When the quantity of any commodity which is brought to market falls short of the effectual demand, all those who are willing to pay . . . cannot be supplied with the quantity which they want . . . Some of them will be willing to give more. A competition will begin among them, and the market price will rise . . . When the quantity brought to market exceeds the effectual demand . . . The market price will sink . . ."
>
> —ADAM SMITH, *THE WEALTH OF NATIONS* (1776)

> "As many more individuals of each species are born than can possibly survive; and as, consequently, there is a frequently recurring struggle for existence, it follows that any being, if it vary however slightly in any manner profitable to itself, under the complex and sometimes varying conditions of life, will have a better chance of surviving."
>
> —CHARLES DARWIN, *ON THE ORIGIN OF SPECIES* (1859)

Spanning the time of the publication of Adam Smith's *The Wealth of Nations* and Charles Darwin's *On the Origin of Species* there emerged two truly original and organic regulatory principles that would come to dominate our understanding of adaptive order in decentralized systems: the invisible hand and natural selection.

These ideas were radically different from the universal ideas of physics that had preceded them and that had provided the liberating

---

OPPOSITE *Economic Curiosity: Once Upon a Time in Shaolin*

empirical soil in which they now grew. What Newton described in the *Principia* was a universe of cold cause and effect:

"It seems to me farther, that these Particles have not only a *Vis inertiae* (want of power to move themselves), but also that they are moved by certain active Principles, such as that of Gravity . . . which causes . . . the Cohesion of Bodies. These Principles I consider . . . as general Laws of Nature."

Whereas for Newton, particles were as simple as possible—possessing only the property of inertia and pushed around by active principles (namely gravity)—for Smith and Darwin, the correct particles were organisms actively seeking scarce resources and regulated by emergent principles generated by the collective activity of populations in competition. Both the invisible hand and natural selection encode the integrated average of purposeful adaptive agents.

The founding of SFI was a very active effort to move beyond theory in which homogeneous elements move in response to universal forces and fields, towards heterogeneous agents responding to emergent incentives and rewards. And the biological sciences and the social and economic sciences are those domains where these features predominate, hence their critical founding role in SFI's history. To this day many are still looking for the keys to insight under nineteenth-century gas lamps (or even eighteenth-century oil lamps). The fact is, our theories are still very far from capturing complex reality.

This meeting is in some sense an effort to assay the status of this effort. How far have we come? How have advances in computer simulation, non-linear dynamics, network theory, and the theory of computation, overcome the significant theoretical challenges of the complex domain? And then there is the monumental question of the application of new economic theory in the evolving economic systems of the planet.

John Maynard Keynes, writing in his "Economic Possibilities for Our Grandchildren" (1930) suggests that "We are suffering . . . from the growing-pains of over-rapid changes, from the painfulness of readjustment between one economic period and another. The increase of technical efficiency has been taking place faster than we can deal with the problem of labour absorption."

Keynes was concerned that rapid change would be incompatible with social equality: "What can we reasonably expect the level of our economic life to be a hundred years hence? What are the economic possibilities for our grandchildren?" And he concluded that the future for the mass of humanity might surprise us: "The course of affairs will simply be that there will be ever larger and larger classes and groups of people from whom problems of economic necessity have been practically removed." This would transpire as long as we showed a greater willingness "to entrust to science the direction of those matters which are properly the concern of science."

It is noteworthy that Keynes was an avid collector of Newton's manuscripts and the author of one of the most insightful biographical essays on Newton ("Newton, the Man") published posthumously in 1946—the year of his death—and whom he described as "one who taught us to think on the lines of cold and untinctured reason" and "that he could reach all the secrets of God and Nature by the pure power of mind—Copernicus and Faustus in one."

For Keynes—like so many who came before him who similarly channeled the Euclidean hopes and dreams of the Enlightenment—that social culmination of the pure lines and curves of the Newtonian paradigm—economies might be tamed by a quantitative social physics.

There is an alternative paradigm that studies the interactions of the economic system with other planetary mechanisms to include the global climate, ecosystems, social life, behavior, and political institutions. This level of integration—moving far beyond the mechanics of the seventeenth century to encompass the regulatory principles of Smith and Darwin—represents the scale of challenge for an SFI program in complexity economics. And we would argue it is a need if we are to ensure the prosperity of all life.

6

CAN COMPLEXITY ECONOMICS SAVE THE WORLD?

Introduction & Talk by Eric D. Beinhocker

This talk is derived from my forthcoming book, *Market Humanism*, co-authored with Nick Hanauer. The thesis of the book is that in order to change the economic system—to create a system that is more inclusive, prosperous, and sustainable than the one we have today—we have to change the ideas that built that system. An understanding of human beings as the social, cultural, cooperative creatures that we are, and economies as the complex, evolving systems that they are, is central to the new thinking that is required. Much of this work was pioneered and continues to be developed at SFI. In this talk I sketch out where the economic ideas that inform our current economic ideology and system came from, why those ideas have been so damaging, and how new economic thinking could lead to a new ideology that helps us meet our challenges.

ERIC D. BEINHOCKER This is going to be a bit more of a philosophical talk rather than a scientific talk. What I hope to do is raise some issues that will then set up the more scientific discussion that the next speakers and the rest of the afternoon will bring to us. In that spirit, I'm going to start off with a philosophical question, which is, "What is the economy? What kind of a system is it?"

Well, there's a wonderful phrase that the historian Noah Yuval Harari uses to describe human social structures—he calls them

---

OPPOSITE *Economic Curiosity: Leonardo da Vinci's Salvator Mundi imagined orders*.[1] I would argue that the economy doesn't exist in reality. It's a product of our imagination. It's made up by human beings. It exists in our heads. The economy is made out of ideas. And if you look at all of the economic ideas and concepts that we use very commonly, whether it's economic value, money, markets, growth, GDP, wages, or even the idea of a job—all of these things are concepts that were made up at some point in our history. They're not part of the physical world or governed by immutable physical laws. They are human conceptions. So, the economy is a product of our imagination. It's an imagined order. Now, this imagined order has evolved quite dramatically over time. Our economic arrangements have changed a lot from hunter-gatherer days to the advent of agriculture, the Industrial Revolution, and to modern capitalism today. And that imagined order then does have an effect, an impact, on the real physical world. It changes our physical world.

When something happens in the imagined order of the economy, buildings get built, food gets grown and delivered, products get made, and so on. And we've seen the impact of the imagined order of the economy on our physical world at a planetary level, causing a transition from the Holocene to what now is being called the Anthropocene. As this has unfolded, we're now beginning to realize that the imagined order that we have is not succeeding.

A chart from ecological economist Dan O'Neill and collaborators shows the failure of the current system in one concise graph.[2] It indicates that a number of societies have succeeded pretty well socially in providing material and social benefits for their populations, but are breaching a wide set of biophysical boundaries and literally destroying the planet. There is another group of societies having relatively little biophysical impact, but failing on multiple social dimensions. We need to be succeeding on *both* the material and social levels as well as in minimizing our biophysical impact. Sadly, literally no society on Earth is doing that. Our imagined order is not delivering on the two great priorities that it has of delivering human well-being and a sustainable planet.

1 Yuval Noah Harari. 2015. *Sapiens: A Brief History of Humankind*. HarperCollins: New York, NY.

2 See O'Neill, D. W., et al. 2018. "A Good Life for All within Planetary Boundaries." *Nature Sustainability* 1: 88–95, fig. 2.

Now, what's inside this imagined order that organizes our economy? What is responsible for this fundamental failure? Well, I like to think of our economic system as a set of nested Russian dolls (slide 1). We have the economic system, and then within that we have a set of economic and political ideologies—normative ideas about how things should be organized. And then, inside that doll, we've got a set of economic theories, positive explanations about how the system works. And then underneath that there's a set of theories about behavior, about how human beings work, because, after all, economies are made out of people. Finally, deep inside the nested dolls are some ideas about moral philosophy, about what's good and bad and right and wrong.

We need to be succeeding on *both* the material and social levels as well as in minimizing our biophysical impact. Sadly, literally no society on Earth is doing that.

Now we can put some names on the nested Russian dolls to describe the system that we have today. At the top level we can refer to our economic system, at least in the West, as modern market capitalism. Inside that is a dominant economic ideology, often described by historians as neoliberalism. Next, the dominant economic theory for the past many decades is neoclassical economics. Then inside that is an idea of human behavior that can be called *Homo economicus*. And finally, inside that is a moral philosophy of maximizing utilitarianism, which I'll talk a little bit more about in a moment. In looking at the history of how this current thought system came to be, we can see that the set of intellectual ideas in the smaller Russian dolls came to intellectual dominance largely in the nineteenth to mid-twentieth centuries. The ideas then came into political dominance really in the 1970s through 2010s, and then geopolitical dominance during the last several decades as these ideas spread around the world.

There are different variations on this imagined order. Put in crude terms, there's a more conservative version with more markets and less

UNPACKING OUR CURRENT IMAGINED ORDER

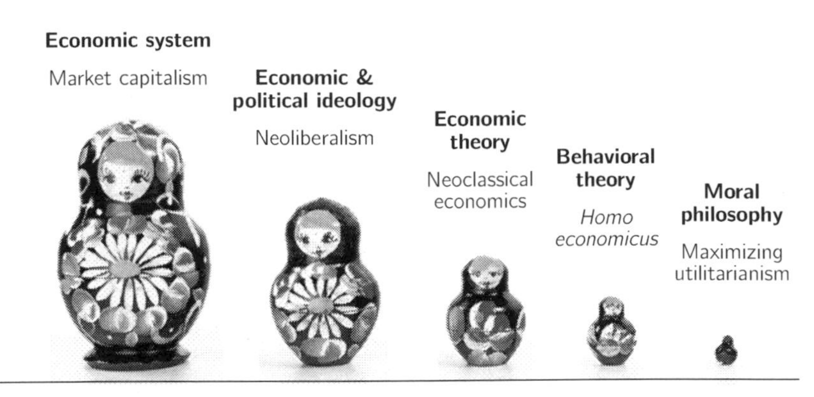

Slide 1

state, and a more progressive version with the inverse. One can think of the US and Singapore as the right doll and Norway and Sweden as the left doll. But these variations are largely built off of the same kind of imagined order and idea system. We have politics in this system and big fights between these left and right branches and arguments about which is better, and so on. But the reality is that both the left and right branches lead to mass extinction. Norway is no more on a path to a sustainable planet that delivers the human and planetary well-being than is Singapore or the US or any other country.

Instead of fighting at the level of policies and politics, we need to be fighting deep down at the root of the ideas that have created this system, from the moral philosophy and the economic theories up to the ideology. If we go back in history and look at why we think about the economy the way we do, the roots of our thinking can be found in long-dead thinkers centuries ago. And in that little doll of moral foundations, there's been some very good scholarship on the huge shift that happened from before Hobbes and Bentham to the era that followed them. I can summarize a thousand years of moral philosophy in a few sentences: pre–Hobbes and Bentham, human nature was viewed as a battle between our desire to be good and our temptations to behave badly, and the gist of moral philosophy and religious faith was that we should treat each other as we want to be treated ourselves—the golden rule—and we should do our best to rein in our worst instincts. Morality was also viewed as a part of character or virtue. There were of course lots of different flavors of this, and debates, but this was the rough framework.

Then Hobbes and Bentham came along and had a radically different idea about human morality: they said that seeking pleasure and avoiding pain is just human nature. It's neither good nor bad itself. It's just who we are. What matters are the consequences or actions that result from our pleasure seeking or pain avoidance. In their view, moral behavior maximizes the most pleasure for the most people and minimizes the most pain. That's the core idea of utilitarianism. Other things like character or intentions and so on don't really matter. What matters are *consequences*, what philosophers called *consequentialism*.

Now, this was quite a major shift in moral thinking and it had some very positive impacts on the development of ideas such as human rights and the radical notion that governments should serve people, not the other way around. But this consequentialist utilitarianism got adopted very heavily in the new field of economics that was just starting to develop around the late eighteenth and early nineteenth centuries. And economists then translated that utilitarian moral framework into a theory of human behavior often referred to as *Homo economicus*—the idea that humans rationally and selfishly pursue their own utility or pleasure—that we recognize today as pseudoscience. It's not empirically valid, this idea that humans are self-regarding pleasure maximizers and rational calculators of utility, and yet it became the dominant behavioral theory of economics and much of social science for over a century. If you look again at the history, you see that the early thinkers were trying to make economics more of a science than a philosophy, more like physics, to be able to create mathematical theories about individual behavior based on this idea. The Nobel laureate Herbert Simon once criticized his colleagues for doing "armchair economics"—dreaming up unrealistic theories from the comfort of their armchairs—and this really was armchair psychology. Basically, these eighteenth- and nineteenth-century philosophers and economists sat in their armchairs with a glass of port or whiskey, swirled it around and said, "I think human beings are pleasure maximizers and pain minimizers and I think they pursue that in a rational way."

They had no evidence, no empirical work. There was nothing scientific about it. It was a purely made-up exercise of philosophy. Yet it came to totally dominate our thinking about human behavior in the social sciences for a couple centuries afterwards. This then led to various normative implications, notably the idea that, crudely put, greed is good and greed creates prosperity. Now, who thought of that idea? Did Adam Smith? This is the message of the "invisible hand," right? Wrong! Adam Smith did not think that at all. It is actually a myth created by Milton Friedman that Adam Smith thought that—but that's another story. Instead, Smith was actually a very subtle thinker about moral philosophy and wrote an entire book on why greed is not good and how what we today would call *empathy* is the key to a successful society. Rather, these ideas that self-interested maximization in markets leads to good social outcomes really comes from a different crowd called the *marginalists*, who came long after Smith, and who gave us modern economic ideas about individuals maximizing their behavior and markets leading to good social outcomes.

This then led to a kind of systems theory of the economy, developed mostly by a chap called Léon Walras, who, again, was trying to make economics more quantitative and more scientific. And so he developed the concept of the economy as an equilibrium system, as an optimizing system that is essentially static and separable from other systems in the social sciences. That led to a set of normative ideas about what's good in the economy, about markets being efficient, governments being less efficient, price equaling value, GDP being a proxy for human welfare, the point of the economy being to maximize GDP growth, and a whole set of other things that then became part of the intellectual, political, and policy environment.

Next came a set of political thinkers who turned those intellectual theories into memes and norms that got operationalized in our politics, policies, and popular culture. A few examples include: Greed is good. Maximizing pleasure from consumption is the goal. A billion acts of selfishness will lead to a prosperous society. The social duty of business is just to maximize its profits. There's no such thing as society, only individuals—that was Margaret Thatcher's famous quote. Markets are efficient; other institutions are not. The value of something to society is its market price. You're paid what you're worth. You have to choose between equity and growth, and the environment is a resource to be exploited.

These are all highly contestable statements. They get debated a lot in our discourse and our politics. But this is the kind of crude, cartoonish interpretation that's come out of economics into the popular culture. These are the memes that were produced by that intellectual system. Now, what I would argue is that if we look at these ideas from a modern scientific standpoint as opposed to an armchair-economics or armchair-psychology standpoint, they actually look pretty absurd. I've always liked this quote from John Stuart Mill: "It often happens that the universal belief of one age of mankind . . . becomes to a subsequent age so palpable an absurdity, that the only difficulty then is to imagine how such a thing can ever have appeared credible."[3] I expect that, someday, looking backwards, we will see just how absurd these ideas were.

If we now fast-forward to SFI, in the modern world, we don't have to be armchair economists and psychologists anymore. We actually have real science and data on a lot of these things and a lot of it's been done here at SFI. So we know that *Homo sapiens* is not like *Homo economicus*, that real human decision-making is inductive and heuristic, modular, emotionally intelligent, and highly, highly social. One of the big revolutions in social science has been a much deeper understanding of humans as social creatures, that we can be both selfish and groupish, reciprocal, cooperative, altruistic, intuitively moral, hierarchical, tribal—a whole set of characteristics about humans as social animals that get left out of much economic analysis.

We also know that the economy is not an equilibrium system. This was one of the key debates in the founding workshop of SFI back in the 1980s, when the economics program was founded. I've always loved this quote from John Holland: "If it's in equilibrium, it must be dead." SFI has been at the forefront of this notion of identifying the economy ontologically as a complex adaptive system, not as an equilibrium system. And this has a whole set of implications for economic behavior, which many of you are very familiar with. And I'm sure Brian Arthur will talk more about this in depth this evening. But it provides a

3 Mill, J. S. 1848. *The Principles of Political Economy.* London: John W. Parker, vol. 1, p. 4.

dramatically different view of how the economy behaves as an emergent system versus the Walrasian view coming out of those Russian dolls that I showed before.

I would also argue that we can start to construct a new conception of economic value based on the kind of ideas that SFI and other complexity scientists have been developing, notably that the economy can be thought of as an order-creating system, that it takes inputs—matter and energy—from low order/high entropy and transforms them via a kind of economic metabolism into all the products and services and social and physical order that we see in the economy. It does this by utilizing knowledge from what can be called physical and social technologies, and obeying the laws of thermodynamics, producing waste, heat, gases, etc., that have consequences for the biophysical environment. This order-creating activity in the economy helps solve human problems. It helps deliver products and services to us that meet our needs and wants, and creates what we think of as value.

Again, this is a very different conception of economic value from what's in the traditional neoclassically based view of the world, integrating ideas from physics and information theory, biology and evolutionary theory. So, if we take this more modern, SFI-rooted perspective of complexity economics, we can start to see the outlines of how we might create a new imagined order that would be much more grounded in science and connected to the reality of the societies and the planet that we all live in.

I'll just briefly sketch out a scientific research agenda that might help us do that: There are huge questions as to how we would construct such a new view, e.g., a deeper understanding of human cooperation, our prosocial behaviors, and moral foundations based, again, on real human behavior and instincts as well as a deeper understanding of how we create order in our physical and social environments, and how that order evolves over time. There was a recent article in *The Atlantic* some of you may have seen, calling for a new science of progress, that we don't have a good theory of human progress.[4] It is true—economics has never

4 Collison, P., and Cowen, T. 2019. "We Need a New Science of Progress." *The Atlantic*. https://www.theatlantic.com/science/archive/2019/07/we-need-new-science-progress/594946/

A NEW IMAGINED ORDER IS POSSIBLE

Moral philosophy
Prosocial behaviors

Behavioral theory
Homo sapiens

Economic theory
Complexity economics

Economic & political ideology
Market humanism

Economic system
The eudaimonic economy

Slide 2

really had a good theory of progress and that's because, I think, progress is fundamentally an evolutionary phenomenon and the equilibrium systems of traditional economics don't evolve. The kind of perspective that SFI brings to the table could create a true science of progress: a deeper understanding of the evolution of economic value and how the economy impacts people's lives. Understanding the economy not as the linear, additive system of traditional economics, but understanding things like business cycles, financial market booms and busts, growth, inflation, inequality, etc., as the emergent phenomena that they are.

Then, finally, how does this system called the economy co-evolve with the biosphere and planetary systems? In the complex-systems view, the environment is not an externality. It's not a system separate from the economy; rather, the economy is fully embedded in that larger biophysical system. And I would argue, just to start closing my talk, that, if we did that, a new thought system, a new set of Russian dolls, is possible (slide 2). We can have a moral philosophy based on our prosocial behaviors, a realistic theory of human decision-making based on empirical science, and an economic theory based on complex systems. I think that could lead to a new set of normative ideas about how the economy does and should work.

While such a new perspective will create new political debates, it won't end politics. It's not going to be, as Keynes predicted in the letter that David was quoting from, that economics will be as boring as dentistry because all the fights will be over. Rather, we'll have a new and more productive set of things to argue about than we do today. In fact, I'm working on a book with my co-author Nick Hanauer on what such a new ideology might look like. We call it *Market Humanism*, and that's something I would look forward to talking about at another time. I think such a new imagined order has the potential then to lead to an economic system that does a much better job at promoting both human and non-human flourishing and helping us sustain the one planet that we have.

Economics has been used to justify a lot of very self-serving behavior. Economics has also been used to justify a lot of behavior that we now know is very damaging to the planet.

Such a thought system could lead to a new set of normative memes in the popular culture and our politics and our policies—including the very old idea that greed is actually bad. It's bad for social structures and social relations. Maximizing human flourishing and non-human flourishing is good. Prosperity comes from solving human problems. We solve human problems by cooperating. A billion acts of cooperation make a prosperous society. Inclusion, fairness, and trust are the foundations of cooperation, and, thus, the foundations of economic growth. The role of government is to foster security and help foster cooperation. The role of markets is to create competitions, to cooperate, to evolve new and better solutions to human problems. The social duty of business is to solve human problems. And the moral duty of humankind is to protect life on Earth.

We could imagine then a *new* imagined order arising out of this new thinking and new economic ideas that could lead to a very different kind of economy than the one we have today, hopefully one that can bring us to social and biophysical sustainability. So, can complexity economics save the world? It has to. Thank you.

AUDIENCE MEMBER Very interesting talk, as always. I love the idea that *Homo sapiens* evolve much faster than other animals because they're able to make up stories that basically create this social cohesion and everything else. Throughout history, that emergence has been hijacked by ideology, by religion, and by capitalism, to some extent. But today we see it being hijacked by social media. So, to what extent are these mechanisms part of the whole system versus disruptors to the system?

E. D. BEINHOCKER You're absolutely right. A key part of the development of human culture was our ability to have shared stories, shared narratives that then foster very large-scale cooperation. So, we can cooperate in very complex ways that are sustained over long periods of time. E. O. Wilson titled one of his great books *The Social Conquest of Earth*. And, really, the social conquest of Earth is a story of stories. Plato said those who tell the stories rule society and he was right. Our stories work best when they connect with something in reality, when they tell us something about reality that can help us make human life better. And my interpretation of the Enlightenment was it created a whole set of mechanisms that connected our stories to reality. Science was one of those, democracy is another, and markets are another mechanism.

But that hasn't stopped human beings from trying to hijack the stories we tell to serve their own power and purposes, not always for the benefit of larger society. So that's a very old battle in human history between the self-serving stories of an elite and the stories that can actually help serve a broader society. Part of what we've seen in our economics is that elites previously used to appeal to gods, to how our ancestors did it, to the natural order, etc., to make credible their stories that justify their power and privilege. Well, over the last several decades, they found a new source of authority: economics. Economics has been used to justify a lot of very self-serving behavior. Economics has also been used to justify a lot of behavior that we now know is very damaging to the planet.

Where social media comes into the picture is it is an incredible mechanism for accelerating the spread of stories, making them go viral. But we know from psychology and cognitive science that the stories that most excite our brains are not the most true or useful; rather, they are the ones that trigger emotions like moral outrage or tribal affinity. By splintering our notion of reality and distorting our stories, social media is doing far more damage to society than just the near-term political stuff. It is really an unwinding of the Enlightenment.

It is essential that we find ways to counter this, reground our stories in science and democracy, and then use that shared reality and stories to reconnect our economic system into delivering better outcomes for both people and planet.

AUDIENCE MEMBER Eric, thanks so much for this. I really liked that article you mentioned on the theory of progress by Tyler Cowen and Patrick Collison, the CEO of Stripe. I can't help but notice that all of the biophysical boundaries that you mentioned were being breached are things for which we don't have property rights. The atmosphere, the oceans, federal lands in California, which are now ablaze. People and companies naturally tend to invest in and preserve things that they own, either for their own use or for trade. Most recently we saw rights extended to the electromagnetic spectrum and, even more recently, tentatively through cap and trade, to the environment. Would market humanism include ascribing an interest in the atmosphere, perhaps a one-seven-billionth share, to the seven billion people on Earth?

E. D. BEINHOCKER That's a very good question. So, we need to put markets in a broader context. If you open most economics textbooks, they start usually with some description of exchange in trade as a description of what the economy is. So you get this impression that the economy is largely about markets. But if you talk to the anthropologists, you get a different story, that before you can have markets and exchange in trade, you need to have stuff—products and services, stone tools, hunting parties, physical and social organization. And that came out of human instincts for cooperation. Markets and even notions of property came much later in human development, and organized markets even later than that. The economy really runs on huge networks of cooperation, and markets are one small slice of that huge network.

In our new book, Nick and I say cooperation is like the dark matter of the economy. You know, it's ninety-eight percent of the mass, but markets are kind of like the bright lights, the visible stars that are, like, two percent of the mass, just like in physics. By having such a narrow focus on markets, we've forgotten about the other ninety-eight percent of what enables cooperative societies to do things like build very complicated products and services, have complex social organizations, and solve very complex problems. To have large-scale cooperation, you need a whole set of social and cultural norms and institutional structures. There's a lot of infrastructure around large-scale cooperation that needs to be built. What markets are good at is creating evolutionary competitions between those structures of cooperation. But markets can be harmful when they actually reduce that cooperation and crowd it out.

And so we have to be careful. The neoclassical answer was almost always: if there's a so-called market failure, the answer is to create property rights. Put a price on it. In some cases that might be the right answer. But in other cases—and Elinor Ostrom did some very famous work on this—creating structures to enhance achieving cooperation may be a better answer. And that may be done through cultural norms, institutions, a whole set of other machinery that humans have evolved over very long, long periods of time. On saving the environment, markets may work well in some contexts and not so well in others. Carbon markets have so far not actually abated very much carbon, whereas old-fashioned regulations (e.g., auto emissions standards, building codes) have had some success. So, I would say don't forget the ninety-eight percent of dark matter; markets are important, but they're only one part of the broader picture.

AUDIENCE MEMBER Thank you. I agree with everything you said. But I think the biggest problem is that capitalism is a loaded coin. Every time we toss that coin, it rewards someone, and someone maybe doesn't do as well, and the result of tossing that coin millions of times is the chart that you showed at the beginning, right? The people who have been able to exploit resources better are able to run away with wealth, while others stay behind, and you repeat the same scenario, and see it also at the individual level, right? So, how do we change that behavior that is leading to some outcomes that, eventually, may lead to extinction?

E. D. BEINHOCKER Again, a big question! First, whether capitalism is a loaded coin or not is a choice we make as a society. Again, capitalism is an imagined order. We can construct capitalism in lots of different ways and we can construct non-capitalist ways of organizing the economy. Saying it's an imagined order doesn't mean just anything goes. We can construct the economy in really lousy ways, too. You know, the Soviet experiment was a pretty terrible way to construct an economy and that failed pretty badly. But one of the myths that has come out of our current economic thinking is somehow that the order we have today is the natural order, that it is an inevitable outcome, part of human nature, and so on.

That really diminishes how much choice we have in constructing an order we want. In order to construct a better order, we have to understand that order in a much deeper and more scientific way than we do today to actually know what's going to improve it. We also have to be humble in this complex system about our ability to forecast things and be able to predict what will actually be better. And we need new mechanisms to experiment our way to the future and harness the forces of evolution in the system to experiment and find better ways of doing it. But at the heart of a system that mixes large-scale human cooperation and uses markets to create these evolutionary competitions is the greatest free lunch in history.

> . . . the neoliberal conception of capitalism has damaged the social contract and thus has actually harmed that dynamic and reduced our capacity to innovate and reduced our capacity to create progress.

Economists like to say there is no free lunch, but we have been living off of a free lunch ever since we started banging rocks together in caves, which is to say that by cooperating we can do things we can't do on our own. We know from history that we can solve ever more complex problems and create this bootstrapping dynamic to create more complex physical objects like this laptop, more complex social structures like our systems of law and money, and so on. That is where economic value comes from. It comes from solving problems. We all got the problem of getting fed today solved by a very complex supply chain that delivered us lunch. That was a huge dance of cooperation. So, we need to think about our system in a way that it is harnessing that free lunch—no pun intended—of cooperation in the most effective and fair ways possible.

AUDIENCE MEMBER So, a comment and then a question. The comment is the story of Israel going from a water-starved country to having a surplus which helps its neighbors and is also a big technology industry. I think this is exactly an example of a social contract at the bottom that led to technological innovation. I just wanted to pass that on to you. The question comes from an experience I had as a computer scientist about twenty or twenty-five years ago. A very famous systems computer scientist said, "People will not take passwords and security seriously until there's a catastrophe." And that was true. So, I might agree with what you have on this slide, but do we need a catastrophe to get to the geopolitical systems understanding this?

E. D. BEINHOCKER I'm very glad you brought up the term "social contract," because we can't create complex, durable networks of cooperation unless the contracts between the people cooperating are fair and inclusive and engender trust and so on. There is a strand of academic literature called *varieties of capitalism* that looks at different ways of organizing market systems and different outcomes that come from that. And one of the clear findings from that empirical work is that societies based on fair social contracts are more innovative, more dynamic, more inclusive in terms of sharing the gains of growth, and create positive networks or positive feedbacks, where fairness and inclusiveness lead to more cooperation in the system, and the ability to march up the ladder of knowledge, solving more and more complex problems.

What Nick and I argue in our new book is that the neoliberal conception of capitalism has damaged the social contract and thus has actually harmed that dynamic and reduced our capacity to innovate and reduced our capacity to create progress. Instead, it has created a system that rewards rent seeking and value extraction rather than value creation. The kind of work done by Sam Bowles and Herb Gintis and others here at SFI is actually hugely informative in understanding what a fair social contract is. You can get good empirical data on what people view as fair in their social arrangements. On your last question, I'll just

say you may be right. Human nature requires catastrophes to jolt us into change—but with some catastrophes, notably climate change, by the time that happens it may be too late.

AUDIENCE MEMBER It was interesting you had a quote from Pinker. His book *Enlightenment Now,* and then the work that the Roeblings had done—there's a lot of evidence that suggests that there's been a lot of progress on a lot of metrics that aren't just growth of GDP. You said you don't know exactly how to define progress, whether it's infant mortality, or choices of leisure, or surfing the internet. And if I say, paraphrasing him, right, he attributes it to science and free markets? So, one question is, why is that not some reconciliation for the current system as creating progress? And then, somewhat related, the one thing that is clear from your picture and other things is that this system leads us to ultimately destroying the planet. And I think the general consensus from a lot of us would be that the reasons for that are bad politics, bad religion, naiveté. But, if you're going to go back to the classical economists' kind of view and say, we're trying to maximize some version of utility or whatever you call it—pleasure, lack of pain—maybe that there's such a discount rate on in the aggregation of these agents that, two, three, four, five generations down the road, the discount rate is so high that it might be rational to move toward that. And I'm not sure if this room disagrees with that, that that would matter in the standpoint of what's the optimal social system.

E. D. BEINHOCKER So, you raise a number of points there. One of my favorite cartoons on the climate-change issue shows a picture of a devastated, post-apocalyptic planet and a couple of people in rags huddled around a fire, one saying to the other, "I think we got the discount rate wrong." That to me kind of sums up the problem, because if that's the question, if that's how things are framed, we're asking the wrong question. We're just framing the problem in an entirely wrong way. Seeing why framing it as a cost-benefit problem and discounting future generations requires a longer conversation, but it is a deeply flawed conceptual framework for tackling the climate problem. Our knowledge about the future is limited. You have fat tails in distributions, irreversibilities; there's a whole number of technical features of that problem that just make it entirely inappropriate for a cost-benefit, discounting framing.

But there's a deeper moral question here. There's a big debate that's been going on in other parts of social science that hasn't quite reached economics yet, but needs to, which goes back to my little moral philosophy tiny doll on the very inside of all of the Russian dolls. We've had this hedonic conception of the moral good that basically the purpose of life is to maximize pleasure, that the idea that the good life is one where we're living it up, consuming a lot, you know, basically it's a big party.

You start thinking of progress in terms of: Do people have good healthcare? Do they have decent housing? Do they have jobs that give them a standard of living and personal dignity? Do people have autonomy and control in their lives?

There's a couple of problems with that view. One is, it's just contestable from a moral philosophical standpoint, but it also actually doesn't fit with the empirical work on moral psychology, when people try to understand what people actually really do think of as the good life. Humans are much more multidimensional than that. Sure, I like to have ice cream. I love ice cream; in fact, it gives me great pleasure. But is the purpose of life to get as much ice cream as I can? No. You know, we all have a much broader set of purposes in our life and this ties more closely to the Aristotelian philosophy of *eudaimonia*, of human flourishing, of the idea of the good life as having multiple dimensions to it, including social connections, purpose, being able to contribute to society, dignity, respect, not the narrow Benthamite, utilitarian economic view of humans as just pleasure seekers and pain avoiders.

Led by the ecological economists, there's now a movement to bring this broader conception of the good into economics and that leads to a very different way of measuring progress and alternatives to things like GDP. You start thinking of progress in terms of: Do people have good healthcare? Do they have decent housing? Do they have jobs that give them a standard of living and personal dignity? Do people have autonomy, security, and control in their lives? You know, a whole set of other dimensions to human well-being than just maximizing pleasure. And that is critical, because the system of maximizing consumption, of maximizing pleasure, is literally eating the planet.

PAULA SABLOFF[5] There are some anthropologists who use the word "cooperation" to talk about feudal lords and serfs working together. Obviously, that doesn't include the tremendous power dimension in their definition. I'm wondering how you are defining and using "cooperation."

E. D. BEINHOCKER Yes, I should always be careful when there are professional anthropologists in the room who know about these things better than I do!

[LAUGHTER]

So, this goes back to the point about social contracts that was raised before. Economists actually kind of assumed power out of the field. If you look in the equations of economists or standard economics, you don't see power in there. Yet power relations matter hugely in our social structures. They help determine whether and how cooperation works or doesn't work, or whether it is non-zero-sum cooperation or simply rent extraction and exploitation. There's actually a lot of interest in this in economics now, and anthropologists and sociologists will be amused to hear there's a lot of work going on to bring power back into economics.

But I think as part of our broader understanding of economic cooperation, a deeper understanding of what kind of power relations lead to an expanding dynamic of more inclusion, more cooperation, more problem solving, delivering good outcomes for more people, and which power dynamics just lead to exploitation and actually long-term collapses in cooperation is an important part of the agenda. In our book we argue that power relationships are inevitable, but fair social

5 SFI External Faculty Fellow.

contracts, enforced by strong institutions, can counter exploitive power, and while total equality in material conditions may be neither possible nor necessarily desirable, we can create what the philosopher Elizabeth Anderson calls *democratic equality*—that we are all equal in our democratic rights and dignity in society and ensuring that our cooperation is voluntary and not exploited.

I see the time remaining has just gone to zero. David is signaling one last question. So, who's going to ask me the stumping question to finish on?

AUDIENCE MEMBER Have you given any consideration to extending the ideas you're talking about to other species, and comparing cooperation in other species with how we do it? And is this trivial or is there some use in math that might inform us in different ways?

E. D. BEINHOCKER No, it's hugely informative. And, again, I'm in a room where there are people who actually do this for a living, who know a lot more about it than I do. We've learned a lot from cross-species comparisons of cooperative social behaviors, because humans aren't the only cooperative species. Also, humans aren't the only species that order their environment, that create order to solve their problems. There are lots of other species that manipulate their environment in various ways to create niches that help them survive and reproduce. I personally learned a ton from reading this cross-species literature and talking to colleagues in it, but, again, it hasn't really penetrated the walls of economics very much. So, in the long list of areas where SFI could contribute, I think bringing this broader cross-species perspective, broader anthropological, archaeological perspective, integrating more psychology, and so on into economics could be hugely productive.

Just to finish off, I think if I had to name one place in the world where the right questions are being asked and where there are capabilities to answer these big questions and push forward the program I've described, it's here at SFI. Thank you.

GAS
PUMP
Unleaded
87
CRISP!

CONSUMERS VS. CITIZENS: SOCIAL INEQUALITY & DEMOCRACY IN A BIG-DATA WORLD

Introduction by William Tracy; Talk by Allison Stanger

Noting that liberal democracy optimizes human flourishing, Allison Stanger began her symposium talk by asserting that "democracy is a complex adaptive system under global siege." The exploitation of Facebook and Google's market-based openness by authoritarian regimes, and predictive analytics that treat individuals as consumers instead of citizens were two prime examples she cited in November 2019. Recent events have only emphasized the salience of Stanger's warning; since the symposium, online discourse has distorted scientific insights about the spread of COVID-19, while foreign governments have used social media to manipulate opinions in advance of the 2020 US election.

Complexity science's focus on nonlinear-interaction effects provides a useful lens for considering the ongoing impact of global technology platforms on democratic institutions. This emphasis on nonlinear interaction effects is also a hallmark of complexity economics. For example, Brian Arthur famously linked positive feedback loops to increasing returns and winner-take-most market outcomes.[1] This phenomenon explains why a small number of technology-platform operators have amassed enormous market power. Another example is the generative mechanisms behind rugged solution landscapes; when

1 Arthur, W. B. 1996. "Increasing Returns and the New World of Business." *Harvard Business Review*, 74(4): 100–109.

OPPOSITE *Economic Curiosity: Gas Station Gourmet*

the result of one decision is a function of other decisions, the mapping between outcomes and complete policies is rugged.[2] Stuart Kauffman, John Miller, José Lobo, Mirta Galesic, and many other SFI researchers have explored the implications of rugged solution landscapes on economics, organizations, technology, and policy. Notably, this ruggedness makes it difficult for our institutions, laws, and norms to adapt to the new modes of communication facilitated by technology platforms.

The impact of technology platforms on liberal democracy depends not only on the platforms themselves, but also on the interactions between those platforms and the relevant institutions, laws, and social norms. In this context, institutions, laws, and social norms are often referred to as "social technologies." The co-evolution of physical and social technologies has continued to be an active area of inquiry within the SFI community since this symposium.[3]

The last fifty years has seen unprecedented growth in the physical technologies that underlie human communication. Due in part to the ruggedness of their solution landscape, key components of the relevant social technologies have not been able to evolve as quickly as others. The interplay of interdependent complex systems running on different timescales is an area of ongoing inquiry at SFI; it was highlighted at the 2020 Flash Workshop on Timescales and Trade-offs.[4] In the case of technology platforms and liberal democracy, the divergent rates of evolution have resulted in social technologies that are maladapted to the physical technologies that currently mediate communication. Many of the threats Allison Stanger articulated in her talk resulted from self-serving

2 The ruggedness of a problem's solution landscape increases as the correlation between adjacent policies decreases. Intuitively, rugged solution landscapes "look" like the jagged mountains surrounding Santa Fe. This stands in sharp contrast to the smooth curves and manifolds that dominated twentieth-century economics textbooks, and the difference has strong implications for the ability of boundedly rational agents to identify optimal solutions. A thorough review of this approach can be found in Chapter 8 of Scott Page's *The Model Thinker* (2018).

3 Farmer, D., F. Markopoulou, E. Beinhocker, and S. Rasmussen. 2020. "Collaborators in Creation." *Aeon*, https://aeon.co/essays/how-social-and-physical-technologies-collaborate-to-create.

4 See http://santafe.edu/timescales.

(or even malevolent) actors exploiting this maladaptation at the expense of democratic institutions and discourse.

There are several burgeoning areas of complexity science that help us think about possible solutions to these problems. Jessica Flack, Mirta Galesic, and other SFI researchers continue to work on collective intelligence and belief dynamics.[5] In contrast to earlier Galton-style work on the "wisdom of crowds,"[6] these newer approaches account for nonlinear interactions and feedback among agents. As such, these areas of complexity can help us imagine how our new communications technologies could be deployed to improve the clarity and inclusivity of our democratic decision-making.

The impact of technology platforms on liberal democracy depends not only on the platforms themselves, but also on the interactions between those platforms and the relevant institutions, laws, and social norms.

Imagining a suite of social technologies that allows our new technology platforms to strengthen democratic institutions is the easy part; actually transitioning to this new suite of social technologies will be much harder. Limitations on feasible transitions will ultimately constrain the achievable set of democracy-supporting social technologies. Emergent engineering is an area of complexity science that helps us think through these types of problems.[7] Championed by David Krakauer, Jessica Flack, Melanie Mitchell, and other SFI researchers, emergent engineering aims to develop engineering tools that account

5 Insights from both collective intelligence and belief dynamics will be highlighted at the planned 2021 ACtioN Symposium on The Complex Brain.

6 See Galton, F. 1907. "Vox Populi." *Nature* 75, 450–451.

7 Jessica Flack and Melanie Mitchell's recent *Aeon* essay provides one useful conceptualization of emergent engineering: Flack, J., and M. Mitchell. 2020. "Uncertain Times." *Aeon*. https://aeon.co/essays/complex-systems-science-allows-us-to-see-new-paths-forward.

for agent adaptation and other interaction effects. Understanding these tools could play a crucial role in helping both citizens and policy makers address the problems Stanger raises in the talk below, and preserve democracy in the twenty-first century.

DAVID C. KRAKAUER Allison Stanger is relatively new to SFI, so some of you won't know her. Issues haven't been raised yet of how optimism around the use of data to perhaps transform social sciences also comes with a cost, and perhaps you will address some of those issues.

ALLISON STANGER Yes, I will.

D. C. KRAKAUER I am very much looking forward to your talk, thank you.

A. STANGER Thank you so much for the invitation to speak to you today. I'm currently out and about doing all sorts of public speaking on my current book on whistleblowers, which was released on September 25, the same day that Nancy Pelosi launched the impeachment investigation. So it's a really great break to test drive on you here today some preliminary ideas from my next book, which is in its early stages. And I think it's going to follow very nicely from the first presentation. It will almost look like we planned it, perhaps.

What I would like to do is put forward a new frame: liberal democracy is a complex adaptive system under global siege. I'm going to try to bring politics and governance back into our discussion, as I want to illuminate this dynamic from a multiplicity of angles, from the vantage point of the United States, of the European Union, of the business–government relationship in a global economy.

Here's a road map for what I want to do. I'm going to focus on that line between business and government in the global economy that has become increasingly porous and has made some of the challenges more difficult. What I want to do very briefly is five things. I want to first ask, is justice a function of regime type? And I'm going to argue that it is, and this is going to make the very notion of justice more complex and nationally circumscribed in a global economy.

Second, we'll look at the increasingly porous line between business and government, which is a product of both privatization and of globalization, and how these have combined to hollow out government. I wrote a previous book called *One Nation Under Contract,* which was about the privatization of all these functions in national security that really changed the way Americans think about what government is and those things that only government can do well. I think Barney Frank got it right when he said that "Government is just another word for the things we do together." That's not a consensus statement in the United States today.

Third, I want to look at some of the implications of statistical machine-learning algorithms and privatized government in a global economy, and suggest that their interaction has contributed to greater social inequality.

Fourth, we'll probe some of the Enlightenment's blind spots, and their problematic ramifications in American political development.

Finally, I want to leave you with some thoughts about the future of liberal democratic citizenship in a big-data world.

Now, each of these could be a talk in and of itself, but I want to throw them out there for you. You can consider this a symphonic effort to generate discussion. I hope we'll have a good discussion.

So, why is justice a function of regime type? Surveillance in the internet age, whether by governments or companies, often relies on algorithmic searches of big data. Statistical machine-learning algorithms are group-based. Liberal democracy, in contrast, is individual-based, in that it is individuals whose rights are the chief focus of constitutional protection.

Algorithmic opacity, which can be the product of trade secrets, expert specialization, or probabilistic design, poses additional challenges for self-government because it by definition abstracts away the individual on which a rights-based regime depends.

So, some challenges for liberal democracy in contrast to illiberal regimes: The distinction between the public and the private spheres is not drawn in the same way, and the individual is not the regime's point of departure. Privacy is routinely sacrificed at the altar of national security and societal goals in authoritarian systems.

For example, China's Google, Baidu, has partnered with the military in the China Brain Project. It involves running deep-learning algorithms over the data Baidu collects about its users. According to an account in *Scientific American*, every Chinese citizen will receive a so-called "citizen score," which will be used to determine who gets scarce resources such as jobs, loans, or even travel visas.[8] China also uses facial recognition currently to monitor its Uighur Muslim minority for law enforcement purposes. China's social credit system is designed to reward prosocial behavior and punish so-called antisocial behavior. That is now in the process of becoming operational.

In *The Human Condition* (1958), Hannah Arendt lamented "the absurd idea of establishing morals as an exact science by focusing on things that are easily measurable or quantifiable." For Arendt, the question is not "whether we are the masters or slaves of our machines, but whether machines still serve the world and its things or if, on the contrary, they and the automatic motion of their processes have begun to rule and even destroy world and things." So the widening digital-age conflict between producers/consumers and citizens reflects a heightened tension between market and liberal democratic/republican values.

This is important because, when you look at what is going to be appropriate or ethically permissible to advance progress, and you're going to abstract away governance, abstract away the type of political regime you're trying to maintain, I think you're going to reach some flawed conclusions. That's easy to do because capitalism and democracy developed contemporaneously and symbiotically. For that reason, it is easy to miss this potential divergence between the products of entrepreneurship and human values, for reasons that were outlined in the last presentation quite nicely. In the Chinese context, deploying citizen scores, deploying racial profiling in utilitarian fashion, may be legitimate, but in a rights-based liberal democracy, such algorithmic discrimination must be illegitimate. You might say that bell curves don't matter for human rights, individuals do. So, justice, I think, cannot be divorced from the political order we're trying to sustain.

I think that everyone in this room would agree that liberal democracy is most likely to optimize the potential for human flourishing.

---

8 Helbing, D., et al. 2017. "Will Democracy Survive Big Data and Artificial Intelligence?" *Scientific American*. https://www.scientificamerican.com/article/will-democracy-survive-big-data-and-artificial-intelligence/.

Maybe there are some people who disagree with that, but I would say that's the case. So where does the appropriate business–government relationship fit into that equation? Or has the business of government simply become business?

The March 2016 standoff between Apple and the FBI illustrates the new potential for conflict between business and government interests that technological change has wrought. Apple refused to help the government unlock the iPhone of Syed Farook, who was charged with killing fourteen in the December 2, 2015, San Bernardino terrorist attack. Since Apple's market is global, it had no interest in complying with the FBI's request, as its foreign customers are unlikely to pay a premium for a smartphone that the US government can access. Yet Apple is also a company headquartered in the United States, and American citizens have an obvious interest in preventing future terrorist attacks.

. . . has the business of government simply become business?

So, multinational corporations based in the United States can face a conflict between their bottom line and the interests of American national security. This same friction animated the decisions of Facebook's senior leadership to downplay Russian interference in the 2016 American elections until a free press forced them to own their self-interested choices. I don't know if you remember this, but Facebook originally adopted a three-pronged optics strategy for weathering the fallout. They first pursued an aggressive lobbying campaign in Washington to shift public anger onto Facebook's rival companies and thwart regulation that could undercut Facebook's market share. Second, they tapped business relationships to frame criticism of Facebook's leadership, all Jewish—Mark Zuckerberg, Sheryl Sandberg, and Joel Kaplan—as anti-Semitic.

Finally, and perhaps most alarmingly, Facebook employed a Republican opposition research firm called Definers Public Affairs to discredit its critics by linking them to the liberal financier George

Soros. This prompted a November 14, 2018, letter from the president of the Open Society Foundations, Soros's philanthropic arm. And I just wanted to read a bit for you here so you can get a sense of this tension. This is from Patrick Gaspard, the president of the OSF:

> *Dear Ms. Sandberg:*
>
> *I was shocked to learn from* The New York Times *that you and your colleagues at Facebook hired a Republican opposition research firm to stir up animus toward George Soros.*
>
> *As you know, there is a concerted right-wing effort the world over to demonize Mr. Soros and his foundations, which I lead—an effort which has contributed to death threats and the delivery of a pipe bomb to Mr. Soros's home. You are no doubt also aware that much of this hateful and blatantly false and anti-Semitic information is spread via Facebook.*
>
> *The notion that your company, at your direction, actively engaged in the same behavior to try to discredit people exercising their First Amendment rights to protest Facebook's role in disseminating vile propaganda is frankly astonishing to me . . . These efforts appear to have been part of a deliberate strategy to distract from the very real accountability problems your company continues to grapple with. This is reprehensible, and an offense to the core values Open Society seeks to advance. But at bottom, this is not about George Soros or the foundations. Your methods threaten the very values underpinning our democracy . . ."*

What was Facebook thinking? Well, they worried that if Facebook implicated Russia still further, Republicans would accuse Facebook of siding with Democrats. And they were also concerned that if they pulled down the Russians' fake pages, those who had been deceived by these ads would be outraged. The one thing they clearly were not thinking about was what was best for American citizens, American democracy, and the rule of law. To get a better sense of the magnitude of this problem, Facebook announced in May 2019 that it had deleted more than thirteen billion fake accounts, a number roughly comparable to the combined 2018 populations of the US, China, and India.

This global nature of the ad market for Google and Facebook represents the greatest challenge for American electoral integrity. It's embedded in the very business model. It's not a bug of the model; it's a feature. So, looking to the future, the prospect of an alliance between authoritarian states and large IT monopolies that would effectively merge corporate and state surveillance—and George Soros has warned about this—could facilitate totalitarian control unlike anything the world has previously seen. This is something that's very interesting because, remember, Silicon Valley's firms first sold themselves to the public as promoters of ideals rather than as profit-seeking companies. Until very recently, Google's mantra was "Don't be evil." And Facebook still defines its mission as "to make the world more open and connected." Apple recently rebranded its retail outlets as town squares.

This global nature of the ad market for Google and Facebook represents the greatest challenge for American electoral integrity. It's embedded in the very business model. It's not a bug of the model; it's a feature.

This sincere news-speak made it easier for consumers unthinkingly to trade their personal data for continued free use of the platform. But this may not continue. What you're really seeing is a friction between what we want as consumers and what we need as citizens.

To see the trade-offs more clearly, it's helpful to return to Hannah Arendt and think about two models of man: bourgeois man and citizen man. For Marx and his heirs, the class struggle between bourgeois man and working man, between the oppressor and the oppressed, is the central dynamic. But Montesquieu, Machiavelli, Montaigne, and the American founders, however, "ransacked the archives of antiquity," as Arendt puts it, to imagine a different model of man for the New

Republic. And the model man of this new system, built on the Roman conception of the public sphere, was the citizen of the Athenian *polis*.

The American Constitution's architects thus drew upon both Greece and Rome in imagining the New Republic of the United States. In this vision, inclusive citizen engagement in American political life is essential for both self-government and human flourishing. While it is a fact that the original vision excluded Blacks and women from citizenship, it's equally true that the same values evolved over time to include all humans of voting age.

When it inadvertently reinforces inequality, statistical machine learning's focus on efficiency may pose problems for liberal democracy. But there's a competing claim that statistical machine learning can reinforce consistency and uphold egalitarian principles by identifying bias. I think this is a really fascinating area for empirical research. Put another way, if injustice tends to be group based, we see inequality aligned along group lines. Does that mean we need to address injustice by focusing on groups rather than the individual? The Black Lives Matter movement is premised on this understanding. Or is it instead the case that race-based data collection will simply reproduce and reify racism on a grander scale? What seems clear is that the assumption or ideology that presents free markets as the solution to all challenges of public life is no longer tenable. Whenever the people allow companies to pursue profit maximization relentlessly in a global market without attention to consequences, producers are unwittingly elevated over citizens. Along parallel lines, whenever citizens allow their personal data to be harvested in exchange for a better deal, consumers are inadvertently elevated over citizens. Yet we need citizens in a democracy.

The legacy of the Enlightenment's blind spot makes it easy to not see this tension and overlook the most unsettling aspects of our present predicament. The dark side of a big-data world, where the virtual and the real, as well as the public and the private, no longer have clear lines of demarcation, has prompted Harvard Professor Steven Pinker to call for a revival of Enlightenment values, both humanist and scientific, to reinvigorate public life and democratic deliberation. Pinker defines such humanism as "the call for maximizing human flourishing." Yet the Enlightenment's association with slavery, colonialism, and domination cause others to define the Enlightenment as the problem rather than the solution. If you've followed popular internet memes, the *OK Boomer* meme is a good example of that latter orientation.

If our aim is reform that enables human flourishing, the key question to ask is, *Who is doing the lion's share of the flourishing in the existing system?* With spiraling economic inequality along racial lines in the United States, the question can no longer go unasked; if we do not acknowledge the Enlightenment's abuses, those who are suspicious of the Enlightenment's hypocrisy are unlikely to see moral questions in the same way. And here I just want to look back on the efforts of once well-respected American scientists in the nineteenth century, such as Charles Pickering and Louis Agassiz, who believed, as Silicon Valley seems to do today, that being able to measure something was synonymous with understanding it.

If our aim is reform that enables human flourishing, the key question to ask is, *Who is doing the lion's share of the flourishing in the existing system?*

Entirely unaware of the influence of their own self-interest and biases on the questions they found interesting, they developed polygenism (the idea that the races were different right from the start as opposed to being that way from genetic degeneration). They did this to explain the racial inequality they observed, which was a comforting way of seeing the institution of slavery as a legitimate response to natural inequality. This problem of self-interested reasoning and scientific rationalization of the status quo was not exclusive to racial fantasy in the United States. Aristotle's *polis* in Athens was the preserve of property owners only. German *Kultur* birthed spectacular art, music, and philosophy—as well as Nazism. Reviewing the Enlightenment's abuses and its triumphs demonstrates the necessity, as Theodor Adorno and Max Horkheimer urged in *Dialectics of Enlightenment*, for the Enlightenment's champions to engage in a bit of critical self-reflection and historical honesty in proposing remedies for the ethical dilemmas humans face.

What I'm really saying here is that human beings—all of us; this is not just some of us—have blind spots that need to be owned if we are to confront contemporary inequality in America with requisite force. And for that we need reason and free inquiry, not militant ideology. The best way to right wrongs, advised Ida B. Wells, is to turn the light of truth on them. So where does this leave us with liberal-democratic citizenship in a big-data world? If we take Wells's admonition to heart, what is happening to democracy when foreign-purchased ads can swing an American election? Money can amplify a message in social media. And, to date, Facebook has shown no interest in whether the money is foreign or American, although this may be in the process of changing.

. . . the sustainability of liberal democracy requires all of us to be citizens first and consumers second.

Facebook's tools of micro-targeting, its machine learning, and its paid promotions are all utterly opaque, and they unwittingly worked in tandem with the Trump White House to elevate lies over truth and wealth over American citizenship in the 2016 elections. I think engineers and corporate leaders alike need to be mindful of this decoupling of profits and profit margins from the common good and democratic consequences. All Western computer scientists should care about the consumers-versus-citizens tension, not only because as citizens they value liberal democracy, but also because the long-term sustainability of the companies for which they work depends on it. When platforms or products appear to undermine human values, brands are tarnished in the free world, sometimes irreparably.

The challenge will be to reclaim the public sphere for the people in democratic deliberation rather than as a locus for self-promotion and manipulation. A necessary condition for meeting that challenge will be to reintroduce political thought—thinking about the optimal role of government to scientific knowledge. Politics has no place in scientific research, yet we know, standing in the shadow of Los Alamos, that ideas can be weaponized. When technological innovation outstrips the capacity of the existing norms and laws, it will take more than science to construct safeguards for human development and re-harness technology and science to the public interest.

The thing to remember here that I think we've forgotten is that there are some things that only government can do well—or let me just put it really simply here. Nobody elected or appointed Silicon Valley, so technocratic solutions over the heads of governments and their peoples will only increase popular cynicism that business elites serve themselves rather than their country. So, put another way, the sustainability of liberal democracy requires all of us to be citizens first and consumers second. Scientists and corporate titans alike who understand the inherent trade-offs between what we want as consumers or producers and what we need as citizens can be critical allies rather than enemies of pluralism and the freedom of the individual. Thank you for your attention.

[APPLAUSE]

AUDIENCE MEMBER Hi. I really liked your raising of the OK Boomer meme and I just wanted you to perhaps talk a little bit about the rise of the . . . I wouldn't necessarily call them marginal, but newly empowered constituencies. For example, OK Boomer, #metoo, the rise of feminism in public discourse—what are your thoughts about that and what role will those play in creating our new society?

A. STANGER Well, you put your finger on an important point. We saw the significance of memes in the last presentation as well. It's important to realize that memes aren't arguments. Memes aren't ideas. They aren't even really crystallizations of ideology. Memes are vessels for the emotions. For that reason, I get very worried when memes are driving our politics. If you want to have engaged citizenship and democratic vitality, the public needs to understand the limitations of memes for reaching consensus on where we want our country to go. Since rewriting our national story is going to involve arguments and public discussion and a good deal of listening on the part of the powerful, I don't like the memes. What you're seeing, though, I really think, is a push to reconsider the history we have written together to date. That's what the OK Boomer meme signifies. For all of us boomers, who have created the mess we're currently in, the young want us to acknowledge certain truths that our rosy view of the past has obscured from view. That's why I raised the point about coming to terms with the Enlightenment's blind spots and some of the crazy science that took place in the nineteenth century in this country and in Europe. Louis Agassiz was at Harvard, he was at the top of his game, and he was espousing racist theories.

But a Manichean view of our past, with a recasting of the heroes and the villains of history, is no solution either. Rather, we need to realize that all of us really possess blind spots reviewing the world and its inhabitants through our own lens. The way to combat that, in my view, is to own the true history. Not the distorted ideological version of history, but these truths as someone like Jill Lepore would define them. And that means you've just got to see that the history of the United States has been told in a way to highlight white accomplishment and obscure white malevolence. I wouldn't go as far as the 1619 Project with *The New York Times*, which traces the American founding to the arrival of the first slaves in North America rather than to the Declaration of Independence, since it distorts at the same time that it reveals truths of great value and importance. American ideals aren't the problem. It's the hypocritical implementation of those ideals that is the tragedy in need of reparations and remedy. Hopefully, we are moving toward owning our history in its full human complexity.

The OK Boomer meme, devised by the young, is a way of forcing the powerful to acknowledge their own limitations. It is young people saying, why should we even listen to you? Don't tell me about the glories of American democracy, because you're destroying the planet and you committed genocide against Native Americans and enslaved Black Americans, and you didn't even give women the right to vote until the twentieth century. All of those things are true, and I think they need to be owned in order for marginalized groups to appreciate the rich contributions of Western civilization to human betterment.

That's a potentially dangerous aspect of the migration of college students to STEM fields; it indirectly undermines appreciation of the value of the humanities. I love STEM. I love being in Santa Fe and discussing these problems with scientists. What's frightening to me about that transformation is that it is producing young people with a completely blinkered understanding of human development. They have no comparative perspective for understanding the contemporary moment. There's no historical perspective for evaluating the contours of political challenges. All of that combines to enable young people to look around them at what's going in the United States and say, #BurnItDown. If they understood how American constitutional democracy compared historically to other political systems or how their opportunities measured up against China, or with their prospects in other regimes around the world, they'd see it very differently. But you can't have an appreciation of nuance without historical and comparative knowledge. That's why there are truths that aren't revealed through measurement, but only through careful study and understanding of politics in time, which is what is required to own the Enlightenment's blind spots.

JIM RUTT[9] Just a comment or announcement. The OK Boomer meme had annoyed the hell out of me. So, yesterday I launched the counter meme #BoomerPride and it's going exponential. As of a couple hours ago, we had a couple of hundred thousand shots. And so all you boomers that are tired of OK Boomer, every time you see it go #BoomerPride.

[LAUGHTER]

AUDIENCE MEMBER I'm very tired of the meme hashtag.

A. STANGER I'm even sorry I brought it up, because all the questions have been on the meme, but it does capture some of the dissatisfaction and discontent in the general public today, especially among the young.

MELANIE MITCHELL You mentioned something about the need to rethink the role of government in science. Can you say a little bit more about what you mean by that and what the trade-offs are?

A. STANGER It's a great question. It's important to realize and it's something that has been blurred by this belief that market solutions are always superior, if you can have them. I mean, that's the mantra. It's

9 SFI Trustee.

a Republican and Democratic mantra in Washington that if you get a market solution and contract out for something, it's going to produce a better outcome. But what that's done over time is erode a sense of those things that only government can do well. I love the private sector. I love free markets. We want companies out there pursuing a healthy bottom line. But, at the same time, if we're interested in what serves all Americans, that kind of dynamic on its own is unsustainable. You really need government and citizen stewardship to uphold the common good.

The line between business and government is especially porous in the United States due to the privatization of most government functions. Donald Trump's complete disdain for federal governance is the logical culmination of trends long in motion. Americans learned to discount the importance of government, and technological innovation has accelerated that trend. What you see is a developing tendency in Silicon Valley that they're not only going to take care of public problems; they're going to in a sense bypass the federal government or replace government. Facebook, for example, has announced it will have its own proto-Supreme Court to oversee content on its platform. That governance mechanism is global, whereas the First Amendment is national. We have a judicial system to do that and there's just an enormous potential conflict of interest there, even though I applaud the experimentation. I'm getting to your question about scientific research, in that the same applies to scientific research. There's a real need for government funding of scientific research precisely to ensure that public challenges, and not simply private sector ones, get properly addressed.

It's not that you can't take money from a variety of sources and still be independent and deliver interesting scientific findings without being swayed by the funding sources. But government in the past has played a critical role in scientific development. And I see a movement increasingly that disturbs me in thinking that government itself is no longer necessary. This is not only true for science; it's also true for education. The hollowing out of government means that questions of common purpose and social inequality fall off the table. I think part of what's before us today is reclaiming that notion—of the public interest, of the common good—and devising solutions that keep that in mind and keep government at the table in some of these discussions, say, on artificial intelligence, for example. And you've written about this, so maybe you have some thoughts.

M. MITCHELL Yeah, not right this minute, but I see what you're saying. I think you may be a little over-romanticizing government as being in the public interest.

A. STANGER No, not at all. No. If you read my last book, you'll see.

M. MITCHELL But, you know, over the course of science at least in the last century, we've seen a lot of science being driven by military prerogatives, sort of what the military wants, you know, that part of the government. And that's really affected science very deeply in good and bad ways. It's not clear that that's the kind of government push on science that you're talking about.

A. STANGER Yeah. I think that's a fair point. You know, we have the internet because of DARPA,[10] so maybe there are alternative models to talk about, but I think that government funding is extraordinarily important. You are right that the money has largely been channeled from the Pentagon to research, and that has its serious limitations.

FARHAN MUSTAFA[11] The first question you posed was the global system being under a global siege, but a lot of the things you are outlining are quite particular to the US.

A. STANGER That's correct.

F. MUSTAFA Even the global companies you're talking about, the opposition they're facing in other countries are actually largely governmental. So, what is it particularly about the American society that the liberal democracy is taking this . . . if you were to call it a left turn, whereas other liberal democracies are at different stages of this spectrum? It's tough to generalize about all these different liberal democracies in one statement.

A. STANGER That's right.

---

10 The Defense Advanced Research Projects Agency (DARPA) is an agency of the United States Department of Defense.

11 SFI ACtioN member and Managing Director, Head of Investment Risk Management, and Head of Quantitative Research, ClearBridge Investments.

F. MUSTAFA But this discussion is particularly, I think, fairly American-centric. Can you maybe offer a few comments on why you think that is and what does the global scene look like for liberal democracy?

A. STANGER Yes. That's an excellent question. You're right. I was making it American-centric because I started with an argument about justice being meaningful within a particular political system when I made the comparison with China and the United States. But you're right. Obviously, there are other liberal democracies. They're our allies, and there are varieties of capitalism. My focus was on the global siege of liberal democracy, and that's very real.

I mean, if you look at what's happening with the Trump White House today, that is very much a case of what I would call rule-of-law capitalism versus crony capitalism. The president's allies are oligarchs. And if you're interested in human flourishing, if you're interested in continuing to support small business, entrepreneurship, and foreign investment, you're going to want rule-of-law capitalism, not crony capitalism. So, what you're really seeing—and you see this with the whistleblower protection that I've studied in my last book—the impeachment hearings were brought about by a whistleblower from the intelligence community. The European Union has just passed expansive whistleblower protection legislation, which makes it very possible that you could see additional information coming from Europe on financial flows that could affect Trump's position in the United States, because crony capitalism is a global endeavor.

What I see is a global struggle between rule of law, liberal democracy, and illiberal regimes, often run by oligarchs whose interests are served by crony capitalism. The thing that I'm worried about most are minority rights. Maybe we're no longer going to care about democracy in the United States so long as people who have extraordinary wealth and power keep that wealth and power. So there is this problem: the people at the top of the economic hierarchy have a vested interest in the status quo continuing, but it's on a track to something very undemocratic and all the more plutocratic. That has all kinds of consequences for inequality in the system, which will only grow if we don't take steps to reinforce the rule of law.

The rule of law basically says that the leaders don't get to tell you what the truth is. You have an independent judiciary, you can appeal to that judiciary. You're equal before the law. And a lot of other good things follow from that. There are a lot of other ways to order political societies that could be congruent with wealth and people keeping their fortunes. Maybe if we were interested in not pursuing growth at the exclusion of all other values, maybe that would slow things down a little bit and could have positive environmental consequences. But science is not going to flourish without open societies and freedom of thought and expression. Plutocracy is not a regime that celebrates equality or the subversive ideas that fuel scientific progress.

There really is a global struggle underway here and it turns on politics and governance. And if you leave that out of the economic equation, I think you're going to miss some really important things, as well as an opportunity to build a better world in which self-interest is not the supreme human value, the new paradigm in the making that Eric's presentation so persuasively outlined. But that's a great question. Thank you for it.

6

ERGODICITY ECONOMICS

Talk by Ole Peters

WILLIAM TRACY For this part of the afternoon, we're going to shift gears a little bit and move from the ontological perspective that really defined the first set of talks to a more epistemological perspective, getting down into some of the details. Ole Peters began this work or has been involved in this work with Murray Gell-Mann for some time. His talk is titled "Ergodicity Economics." Ole is both a member of the London Mathematical Laboratory and the Santa Fe Institute. With that, I'll turn it over to you, Ole.

OLE PETERS Thank you very much, Will. So, I'll be talking about ergodicity economics. I will start with a game that many of you will know, but it's still the best way to introduce this topic that I know of.

The game is the following: It's a simple multiplicative game. You toss a coin. If it shows heads, you win fifty percent of your wealth. If it shows tails, you lose forty percent of your wealth. You play this game once a minute, so you're tossing a coin once a minute. As you're playing, your wealth follows some sort of a random trajectory. I could ask this room who would like to play this game, and I'll actually do it. Who wants to play this game?

AUDIENCE MEMBER The people in the other room.

[LAUGHTER]

O. PETERS That's good. This is a sign of success. I've asked this question many times and by now almost no one wants to play anymore.

OPPOSITE *Economic Curiosity: Ambergris*

There are good reasons why you might want to play the game, but I'm glad you don't.

Okay. Let's look at one sequence over sixty minutes, so we were playing this game for an hour. You get some kind of a random trajectory, as you would expect. What do you do with a random trajectory? Well, it doesn't really tell you much about the system, for example, whether this is really a game worth playing or not. It's just some noise.

In 1654, Fermat and Pascal came up with a recipe for treating situations like that. And they said, "Oh, just imagine everything that could possibly happen." So, all the possible trajectories over the sixty minutes and an average over them. And if that average looks good to you—in this case, if it goes up, if you're winning—then that's a good game. And we'll call this average the *expectation value*.

So how do you construct this thing? You construct it by just running many, many, many of these trajectories—infinitely many. These are also called an *ensemble* or the ensemble of possible trajectories. So you'd run ten, you'd run twenty, you get more and more colorful lines, and then you start averaging. So at each moment of time you take the average over all of these colorful lines and you end up with some sort of slightly less noisy trajectory. If you average over more trajectories, the noise goes away and you'll find what you'd probably expected to find, which is exponential growth in the expected wealth coming out of this game.

The conclusion from this perspective is that the game is worth playing, on average, but not everyone here wanted to play it. The question is, why is that? Why didn't you want to play this game? The classic solution to this question is called *expected utility theory*. It starts from that observation that not everyone is optimizing the expected wealth or changes in expected wealth. And it introduces a new concept, which is the usefulness of wealth: actually, your dollars don't matter. It's what they're worth to you that matters. And it's the expectation value of that that you will optimize. So, in technical terms, you optimize the expected changes in the utility function.

So, in pictures you have some random wealth trajectory from this game and you translate that through a nonlinear transformation into some trajectory of your utility. And then you take expectations. In this case, if your utility is logarithmic, you'll see the expectation value of wealth going up and the expectation value of utility going down. Problem solved. If your utility is logarithmic, you won't play this game. So, this is the classic answer to the question of why you wouldn't play the game.

It's a bit ad hoc. We noticed behaviorally, just by observation, that people don't optimize the expectation value of money. That doesn't work. And then we said, "Well, let's just compute the expectation value of something else," and choose that something else so that it sort of fits with what we see. It also gives you an immediate focus on psychology, not circumstance. So these utility functions are believed to be neurological imprints in your brain, essentially. This is your character. This is who you are.

An alternative solution that we came up with is now known as ergodicity economics. And here are some of the people—Murray Gell-Mann is one of them—who have been working on this paradigm. And it really starts by going back a long time. Looking at what Boltzmann said in the 1870s in the development of statistical mechanics, he asked the question, "What do these expectation values actually mean that we compute for random systems?" When something happens to the expectation value, is that actually also what happens over a long time to the thing that we are observing?

So, let's try it out. Let's take this game and play it for a long time and see what happens. We start again with a sixty-minute trajectory, and then we just extend it. We play for a day instead of an hour and then we play for a week or maybe a year. And what you see is that there are absolutely no fluctuations in this trajectory. So, with probability one, if you play this game long enough, you will observe an exponential decay.

That, of course, contradicts somehow the initial analysis. So, it's a different analysis. It's true that the expected value of your wealth increases exponentially, but it's also true that over time your wealth decreases. This is quite a puzzling thing to get your head around, and it has lots and lots of implications for economic theory. This sort of a situation where each individual trajectory in the long run does something different from the average over a large ensemble is called *non-ergodicity*. So, this is a non-ergodic system.

In our phrasing, we will just say, if it's a long-term losing game, you won't play. So, this is the ergodicity economics explanation for the observation that people wouldn't play this game necessarily. Let's just quickly recap: Maximizing the expectation value of wealth changes is not always a successful model and people don't really do it. And we now have two alternatives. We have the classic solution, expected utility theory—replace money with utility of money. We have an alternative solution, ergodicity economics, which has replaced the expectation value of the operator on this random object with a time average, or time average growth.

The question is, do these two approaches meet somewhere? Can we hook them up? Can we understand something about the original solution with the new solution? And the answer is yes. There is an ergodicity economics interpretation of expected utility theory. This is the key to everything. If you see that the expectation value of a utility function is being optimized, then, in our way of thinking, this is because the changes in the utility function are ergodic. That means that, over time, the individual will then experience the expectation value of these changes.

. . . if it's a long-term losing game, you won't play.

Okay, a different way of saying this—from a 2016 paper with Murray Gell-Mann—is that there are paths, so you can have some sort of a wealth dynamic and an appropriate utility function for that, which is what we call an *ergodicity transformation*. The dynamics is what sets the appropriate utility function, a very different view of human motivation that comes out of this. So, optimizing utility changes over many parallel worlds actually also optimizes wealth itself over time. That's sort of the message.

Now, the two approaches are conceptually completely different, so the question arises, can we think of an experiment to see which one of them works better? Can we get some kind of a discriminating experiment? And this is where a group of neuroscientists from Copenhagen comes in; they designed such an experiment. They said, well, expected utility theory says people optimize the expectation value of utility changes irrespective of dynamics; this doesn't enter into the theory. Ergodicity economics says people optimize over time according to the dynamics—so whatever that means according to the dynamics.

Now, that led to an idea: change the dynamics and see what people do. If they don't care about the dynamics, they will behave the way they behave according to their utility functions, to their neurological imprint. And if they do care, well, then, they will change. So they invited people to come in and gamble and they observed their gambling behavior under different types of dynamics. They invited them for one day, and they gambled under multiplicative dynamics, where your wealth is multiplied throughout repeated gambles. On another day, they gambled under additive wealth dynamics, where something is added or subtracted from your wealth repeatedly over many rounds of some gamble.

Then they inferred the utility functions that people were using more or less, based on observed behavior, and compared that to the predictions from ergodicity economics. To do this, they fitted this function here, which is a q-logarithm, and it has the property that it can be both a logarithm and a linear function. And the ergodicity economics prediction for the two setups is that, under additive dynamics, you should see a linear function which corresponds to this parameter η being zero. For multiplicative dynamics, it should be a logarithm, which corresponds to η being equal to one.

Okay. These are the predictions of ergodicity economics: additive, you should see η equals zero; multiplicative, one. So, the observations are these: you can fit these parameters. You get some distribution of the parameters. This is sort of an aggregate picture of the situation averaged over all participants in this case. And you see that, indeed, on the additive day—this is a Bayesian posterior distribution of the probability or likelihood for this parameter being whatever it is, so zero—this parameter has sort of peaked around zero. And on the multiplicative day it's peaked around one.

You can break this up into individual people and see how they behave under these two different dynamics. And you can see that everyone changes their behavior more or less significantly—maybe not this one—according to the dynamics that you present them with. And they all do it in the way, at least qualitatively, that ergodicity economics predicted.

Okay. So, this is actually really big news. I was in Copenhagen recently. I visited this group of neuroscientists and they were very excited about it. They planned many, many more experiments to see how far this holds and what they can do with it. And the key point is that we are much more adaptable than we thought we were. So, these neural guys have, in a sense, inherited the models from the economists that gave them a structure that was believed to be stable, the utility function. This experiment shows that that is actually not stable, and that we can predict in what way it's not stable and how it changes.

This means that aspects of our behavior that we thought to be neurologically imprinted in our brains are actually quite flexible and we can relearn them quickly. And that's kind of *wow*, right? Because it means we can do more; we're more flexible in our brains than we thought. It sort of opens the doors to many interesting questions in neuroscience, maybe even to ideas for therapies for pathological behavior. If we can retrain more of our brain than we thought, that would be really cool.

All right. I'll stop here talking about ergodicity economics. The story is that it gives you an alternative form of economic theory. So, we can kind of arrive at neoclassical economics, just from a very different starting point. And it then addresses all of these key problems or interesting applications that you would otherwise often run into trouble with.

[APPLAUSE]

MELANIE MITCHELL Thanks for that talk. It was great. This question isn't completely well-formed, but in the first talk Eric Beinhocker put up a slide that had the scientific research agenda for complexity economics, and it was dealing with kind of very high-level ideas about cooperation and social issues and so on. And I'm just wondering where you think these ideas are going to make the most impact in rethinking those kinds of things.

O. PETERS Yeah, we were actually discussing this just before I came on. I mean, the nature of this work is really to go back to a quite basic, simple mathematics. That sounds a bit silly to say it's simple mathematics, but it's a toolkit that was not available—because it hadn't been developed—to the people who laid the foundations for economic theory. It's really, really fruitful to take this stuff that was developed 200 or 300 years later and ask the same questions that were asked at the beginning and just develop from there.

So it's something very foundational, but it's quite amazing how much it changes your perspective. You mentioned cooperation. We have a paper on cooperation, just how optimizing long-time growth directly generates cooperation, by nothing but risk management, fluctuation reduction, and so on. So, yeah, maybe we'll leave it there.

> This means that aspects of our behavior that we thought to be neurologically imprinted in our brains are actually quite flexible and we can relearn them quickly.

ERIC D. BEINHOCKER Just to chime in on that, because it's a great question. I think it very fundamentally is a part of that broader research agenda. What you're doing is tremendously important because, if you take an evolutionary view of the economy, you have this combinatorial opening of the kind of space of products and services and you can't pre-state the sort of phase space of the economy. In such a world, evolutionarily, maximum utility theory wouldn't be a very good strategy. It wouldn't work very well. There's an interesting hypothesis and it's going to lead into my question. Have you explored this with psychologists and others that, in that kind of real-world evolutionary environment that the behavioral strategies, that ergodicity economics, points to, are those the ones that kind of make sense in an evolutionary perspective? It sounds like you're seeing them in the lab in this Copenhagen work.

O. PETERS This is exactly the idea of the Copenhagen experiment. We are changing the dynamics, which is a thing that's difficult to conceptualize, which is why we use very simple dynamics. We just use multiplicative—everyone knows that that's sort of basic stock market models 101—and we use additive, which is a really strange model, where you toss a coin and you always win $1,000 or lose $1,000. The crazy thing there is, for instance, you can go bankrupt and you'd just keep tossing this coin.

So, both have unrealistic elements, but we can do more general things. We can tweak the dynamics in myriad ways. And this is the idea for the future experiments to see, really, if we impose totally crazy dynamics, do people still behave in a way that evolution would dictate? So it's precisely in that line of thinking. Yes.

BLAKE LEBARON I've got sort of a friendly comment and an unfriendly comment.

[LAUGHTER]

My friendly comment is, I think this is part of one of the critical behavioral things out there—that people confuse arithmetic averages, geometric averages. Is it multiplicative? Should I be looking at sums of values or sums of logs of values? Totally with you on that. And your experiments are very nice on that.

Any attitude that mathematical finance has not paid attention to this I think is totally wrong. And one example of this is, I actually give my master's students almost your simulation to do every year to say, "What's the probability that I'm less than the mean, you know, like, seventy-five percent or something like that, you know?"

And then there's a much larger literature. I would cite SFI work on this. Larry Blume, who was a director of the economics program, has a great paper with David Easley on evolutionary financial markets.[1] And of course the survivability there basically turns out to be a mathematical distance from the log functions or essentially growth-maximizing stuff.

So, there's this sort of giant stuff out there. There's even a pop book on it called *Fortune's Formula*. People like to debate about the last guy on *Jeopardy*. So, you know, this world is a pretty big space out there thinking about growth rates and what's going on. I actually think obviously about evolution a lot and it's much more complicated, but in a simple evolutionary stage, I'd say Blume and Easley. But, even in the SFI lore, there's actually a lot of work on it.

O. PETERS Yeah. I'm sorry if I gave the wrong impression there. Of course this has been discovered over and over again under various guises and different terms and different languages and so on. If there is anything

1 Blume, L. and D. Easley. 1992. "Evolution and Market Behavior." *Journal of Economic Theory* 58: 9–40.

unique about what we're doing, it's this focus on the ergodicity question. So just say this is completely real. It's as fundamental as it gets.

When you go back to the nineteenth century and look at what happened in the development of statistical mechanics, there were really these two views of probability that diverged from each other. There was Maxwell's initial view, which was the ensemble view, in essence, that a probability tells you something about the relative frequency of some occurrence in an ensemble of imagined systems. And then there's the Boltzmann view, which is the temporal story. If you keep doing something, how often does it happen over time? And the idea that these two things can be really different.

So, that's number one. That may be a helpful way to conceptualize this. The other thing I have to say is finance is light-years ahead of economics. So, yes, in finance you have things like the Kelly criterion. You see these kinds of thoughts popping up all over the literature.

Even in mathematics, there's a book from 1870 by Whitworth, who discusses exactly these things—the difference between an additive average and a multiplicative average.[2] The ideas are old. The question is, how do you embed them in the conceptual space? We are embedding them in this space of ensembles versus time. Yes.

AUDIENCE MEMBER I was wondering if you took the initial experiment, which is, I guess, a fifty-fifty, win fifty percent or lose forty percent, and you've tweaked it to a real-world issue. If you brought the winnings down on average to hit the same expectation, but instead the big loss was at a low probability, that's like a selling options/tail-risk kind of problem. I think many people play that game. Were there any interesting implications if you were to apply this technique or this methodology to that experiment? Is there anything over time conditional on having seen the tail event that people alter their behavior or anything like that?

O. PETERS Yeah, I mean in the end that's just tweaking the parameters of that game. I don't think you'll see something fundamentally different. You will see this difference between what happens to the expectation value and what happens in the long run to an individual trajectory. I think what

2 William Allen Whitworth, *Choice and Chance.*

is quite fun is thinking about portfolios, in essence. So, you could ask what happens over time to a real-world aggregate, a real-world sum.

If you have a hundred of these geometric motions or whatever they are, initially they behave like an ensemble. Then, as they begin to explore the real, full spaces of their possibilities, they start to behave like one system. They become dominated by the largest term in the sum. And this is where this connection comes in to spin glass theory. And that's a lot of fun. In terms of just introducing catastrophic events, it's quite straightforward, but I don't think it gives anything fundamentally different.

FARHAN MUSTAFA[3] You were talking about the use of this stuff in finance. Just putting two tools out there and maybe if you can opine on their relevance to the two frameworks. Monte Carlo simulation, where you take a certain set of assumptions and you run a thousand, a million, ten million iterations, parallel worlds, and look for a distribution versus a decision tree where you go down a certain branch and you can lop off the branch or something like that.

The first seems like an expected utility tool. The latter seems a little bit more—there is dependency that has conditionality on what has come before. Is one more realistically useful than the other? How would you help finance practitioners think about that distinction?

O. PETERS The question is what do you do with the output of those models, right? It is very easy to fool yourself with Monte Carlo simulations, because often the whole point of this is to generate all these possible outcomes. And then you have to ask the meaning question: what do we make of this? So, there's this fan of possibilities, but how do we interpret that? So you sort of get stuck there sometimes.

Of course, there's also a Monte Carlo simulation to just take one trajectory and run this for a very, very long time and see what time does to your trajectory. So, if that's an option, that's nice. I think there's something that we haven't really explored as much as we should have, and it's the relationship between local and global. There's something very fun. You can look at the expectation values of these ergodicity transformations over a short time. You don't need to run something for

3 SFI ACtioN member and Managing Director, Head of Investment Risk Management, Head of Quantitative Research, ClearBridge Investments.

a long time to know what would be optimal in the long-time horizon. So you can use your multiple-world Monte Carlo simulation if you use it on the correctly transformed variable to find out what hypothetically would happen in the long run. That's a really neat thing that I think we should do more on.

W. TRACY In the interest of time, let's thank Ole once again.

[APPLAUSE]

O. PETERS Thank you.

4

FORMAL MARKETS AND INFORMAL NETWORKS

Introduction & Talk by Matthew O. Jackson

One of the many prescient questions that Ken Arrow asked was, “Does the market (or, for that matter, the large, efficient, bureaucratic state) destroy social links that have positive implications for efficiency?”[1] Robert Putnam’s popular book *Bowling Alone* suggested that the answer is “Yes.” Still, we have little empirical evidence that bears directly on this question, and we lack in theory about relevant interacting complex systems.

To anyone who studies social and economic networks, it is clear that social structures are fundamental to understanding what people believe and how they behave. Thus, it is not surprising that networks have played prominent and recurring roles in SFI’s long history of studying complex systems. Yet, we generally study one “system” at a time.

The studies that I discuss in this presentation concern the interaction of two systems in India that I have been researching for fifteen years, as part of a team from MIT and Stanford. It began in a conversation that I had with Abhijit Banerjee in 2005 about how social structures impact economic systems. I was lamenting the fact that we had lots of studies of social networks, and lots of studies of markets, but very little empirical evidence of the impact of networks on economic outcomes. He remarked that Esther Duflo was in touch with a bank in Karnataka that was offering microfinance, and trying to figure out how best to spread

1 From Arrow, K. J. 1999. “Observations on Social Capital,” 5. In *Social Capital: A Multifaceted Perspective*, eds. Dasgupta, P. and I. Serageldin. Washington, DC: The World Bank.

OPPOSITE *Economic Curiosity: Ten Thousand Bitcoin for a Couple of Pizzas* information among a population. From that conversation emerged a project in which we found that understanding people's positions in the village social networks was essential in predicting eventual participation in the microfinance loan program. Together with Arun Chandrasekar (originally a research assistant on the project) this first part of the project took some time to complete and was eventually published as a four-author article in 2013.

It is important to emphasize that social structures and economic systems influence each other in ways that have major impacts on the functionality of both. This is where we come to Ken Arrow's question above. The presence of formal loans in these rural areas turns out to change the social structure. It was serendipitous that we got to measure this. The original study was planned over seventy-five different villages that the bank was entering. However, the financial crisis halted their lending after they had entered some, but not all, of the villages. By revisiting all of the villages, and re-measuring the social networks, we could see how the presence of microfinance had changed the social structures. We found the magnitude and extent of the changes to be surprisingly large. Seeing how networks changed in response to the availability of microfinance led us to develop a new model of network formation that accounts for how incentives, to form and maintain social relationships, depend on the access of some people to formal credit markets. Details are discussed in the presentation, as well as an associated paper.

This is one of many examples of interacting complex systems and how studying them in combination can lead to new insights into how each of them functions.

A takeaway—for both researchers and practitioners—is that there can be strong and symbiotic relationships between many complex systems. It is imperative to account for how any system is influenced by other systems, and feeds back to change those systems. Of course, one can view interacting complex systems as one giant complex system, but it is useful to view them as distinct entities that interact with each other. This is true since the interacting systems often have distinct purposes and structures. Yet, studying one of these at a time misses fundamental issues.

In terms of complexity economics, this perspective suggests a need for increasingly widening scope of inquiry. One does not have to look far for important examples, as our world is facing many dramatic instances of such interacting complex systems, where the economic system is just one piece. For example, COVID-19 involves a human-interaction contagion network, interacting with an economic production and supply network, as well as education systems, governments, and eventually a financial network. Policymakers who make decisions one setting at a time will make substantial errors. They are faced with the fact that widespread lockdowns have an enormous economic footprint, and yet we are developing theory and harvesting data on the fly to guide them on optimal policies for these interacting systems.

This is just one example of interacting complex systems. There are numerous others. For example, one cannot understand networks of international conflict and war without understanding international trade, and one cannot understand climate change without understanding the differing economic and political interests across the world, and the lack of an overlaid international organization.

Further Reading

Abhijit Banerjee, Arun Chandrasekhar, Esther Duflo, and Matthew O. Jackson, "The Diffusion of Microfinance," *Science*, Vol. 341 no. 6144.

Abhijit Banerjee, Emily Breza, Arun Chandrasekhar, Esther Duflo, Matthew O. Jackson, Cynthia Kinnan, "Changes in Social Network Structure in Response to Exposure to Formal Credit Markets," SSRN preprint: 3245656, http://ssrn.com/abstract=3245656

Matthew O. Jackson, *The Human Network*, Pantheon Press 2019.

WILLIAM TRACY Networks have already been applied in a lot of ways to how we think about complex economic systems, but there are still a great deal of gains to be had. Our next speaker, Matt Jackson, from Stanford and SFI, is really working at the forefront of that area. I'm happy to turn it over to him for a talk on formal markets and informal networks.

MATTHEW O. JACKSON Thank you very much for having me here. As was just mentioned, the set of application areas in economics where we're aware that lots of structure and the shape of interactions make a big difference has grown quite exponentially in the last two decades. In particular, there are lots of examples now where we really have gone beyond sort of formal anonymous markets to think about how the shape of the relationships actually impacts the outcomes.

And this is a very partial list, but I think you're seeing this perspective permeate economics these days. In particular, I'll talk a little bit about finance and how we understand how financial systems now are making the world better off, but also putting it at greater risk, and trying to understand how I think in terms of framing this talk, systems are often built for one purpose, but they serve multiple purposes.

In particular, when we think about our networks—you know, our friendships, our colleagues—they're shaped for certain purposes, but they end up having impacts in other areas. One example is labor and labor markets. We get our jobs through referrals, but we don't form those friendships often for referrals. Even networks like LinkedIn serve lots of purposes beyond just getting people referrals to jobs.

Understanding how these systems interact is very important. And in terms of understanding complex systems from SFI's perspective, often we think of a particular system at a time. What I want to point out here is that these systems interact in fairly complex ways, and they interact in ways that can be unexpected. And what I'm going to do is talk a bit about some projects that I've undertaken with Abhijit Banerjee, Arun Chandrasekhar, and Esther Duflo.

I'll give you two snapshots of this project. This is actually timely; Abhijit and Esther won the Nobel Prize in economics about three weeks ago. So, they've been recognized for doing development economics writ large. This project started out as a fairly specific question: trying to understand how people get information about formal market opportunities. And what it ended up being was more than that. I'll explain to you the starting point and then the end point of the project.

In particular, there are two main points that we'll go through. One, how informal networks make a difference in spreading information about formal market opportunities. In this particular case, we were interested in spreading microfinance loans among a population of poor people in rural India. We were trying to get information about these kinds of programs out to people, and it turns out that the informal networks were the ways to get people informed. So those networks were important in making sure that that market could work.

The second part is, actually, once the loans were undertaken and people started getting microfinance, it turned out to change their social network structure. There was a feedback effect in the fact that now they had money that changed the way they interacted with each other—and actually changed the nature of those interactions. The first part of the project was just figuring out how the networks help the markets, and then the second part of the project turns out to be how the networks actually change in response to those markets. So the markets made a difference eventually.

I'll just go through this project. It's just a microcosm of one of these applications. So the empirical background was, we've partnered with a bank called BSS. They were interested in figuring out how they could get better participation in their microfinance loan programs. These are Grameen-style loans given to women aged eighteen to fifty years old. Their loans were on the order of roughly $200 for a fifty-week period at about a thirty percent interest rate. The idea here was they were going in and offering these non-collateralized loans. In some villages they'd get very high take-up and in other villages they'd get no take-up. They were trying to figure out how they could better spread news about this. They're doing it by word of mouth.

There were seventy-five villages that they were planning on entering. Eventually, in terms of the study to try and understand how networks were changed, it turns out that the bank entered forty-three of these villages. Before the bank entered the villages, we mapped out all the networks. They came into the villages, started offering microfinance, and we traced out how information about microfinance spread and how the take-up of these loans spread.

Serendipitously, in 2009 and '10, after the financial crisis, the bank pulled back on their loan program and stopped lending. They'd entered forty-three of the villages but hadn't entered the other thirty-two. Afterwards, we went back into the villages, and we could see how the villages that got the loans differed from the villages that didn't. We can do a comparison between these villages that got loans and the ones that didn't, and see how the network structures changed in response to that.

Okay. All of this took place in rural Karnataka. This is more or less about a hundred-kilometer band around Bangalore. This is a typical village: They're fairly poor villages, about $1 to $3 per capita income per day—various agricultural things that they're involved in for their productivity. I don't know if you've been following the air quality in Delhi, but this is sort of a typical December picture. They burn pretty much whatever they can get their hands on for warmth. Here you see cow pies on the roofs. The air quality tends to be pretty heavy. These are villages where people don't have access to formal loans, and then this was a program where the bank was coming in and offering these loans.

So, we went into each village and mapped out the social networks in those villages. This is a typical village and each one of those little teeny dots is an adult. They're grouped into households. The households often have several generations of adults in them. And then we asked them a series of questions. We have twelve different networks for each village. One of the questions we asked them was, "If you needed to borrow fifty rupees for a day, who would you go to?" Fifty rupees is about a dollar, so it's about a day's wage. Then there's an arrow pointing to somebody else if that person pointed to that other person.

From that we can start to build a network. We have borrowing. We have: Who do you go to temple with? Who do you go to for important advice? Who comes to you to borrow kerosene and rice? Kerosene is used for heating and cooking. Who do you go to in an emergency for medical help? There's a series of different relationships, and from this we can see which households interact with which other households. There's a structure there.

This is a different way of picturing the same networks. This is one of the villages, Village 26, and this is the kerosene and rice network. I think of the kerosene and rice as sort of the backbone networks in these villages. This is one of the most heavily transacted networks. In particular, this one now is colored by caste. Each one of the little dots is now a household, and there's a link between two households if they borrowed kerosene or rice from each other.

The blues are what are known as scheduled caste and scheduled tribes. These are the ones that are relatively disadvantaged and considered for affirmative action by the Indian government, whereas the reds are the sort of forward caste and then the otherwise backward caste. So they're relatively advantaged castes. You can begin to see this is just drawn by a standard spring algorithm that tries to pull nodes together if they're connected and push them apart if they're not. It uncovers this idea that there's segregation in these villages, heavy segregation by caste, which shouldn't be surprising. But that means if information goes into one part of the network it doesn't necessarily spread to another part of the network. These networks are highly insular, and trying to get information out in the networks can depend heavily on this kind of structure.

Okay. The bank's strategy in going into these villages was to basically look for well-connected people. Their idea was, we want to get the news out. We've got to find the best-connected people in the village, tell them that we're coming in to offer our loans, tell them to spread the news, and then we'll see what happens. So they went in and they looked for shopkeepers, self-help group leaders, and teachers. Those were the categories of people that they thought would be well connected. Sometimes those people might turn out to be somewhere on the periphery. Other times they might turn out to be fairly central.

One hypothesis is, if they hit the highly central people, then they're going to get a lot of spread of information. They're going to get good take-up and participation in microfinance, and other times they won't. So that was one of the hypotheses that we were investigating. There's a long history of trying to understand the right way of measuring who's influential and who's good for spreading information, who the sort of powerful nodes are. And I'll spend a little time on that.

This goes back to Georg Simmel in the 1900s, but there's sort of a long history of this—of trying to identify the most influential nodes in a network. And I'll just take you through a few ideas behind this to try and understand what actually happened in these villages.

The most basic thing is just a popularity measure, right? Usually when we think of influence, we think of who has the most followers on Twitter, or who has the most friends on Facebook, so there's just some measure of just raw numbers. This is known usually as degree centrality. A degree of a person is just how many connections they have. Here, if you are looking at these two networks, these two nodes that have the most connections seem to be the places you'd want to inject information if you wanted to spread it in those networks.

Okay, that's great. So, what's wrong with degree centrality? Degree centrality does not take into account broader position in the network and how well connected your friends are. For instance, these two Twos both have two friends, but it's clear that this one is connected to a Seven and a Six, this one being connected to two Twos. There's a sense in which the one with the Seven and Six is better connected. It's also more centrally connected somehow in this picture. So, we want some notion that captures the idea that people not only have certain numbers of connections but have well-connected friends.

The way that people usually deal with this is something that's known as eigenvector centrality. The concept is actually quite simple and it works out nicely mathematically. The idea is that my centrality is not proportional to how many friends I have, but is proportional to the total centrality of my friends. If I have more central friends, that's better. Having a lot of friends can be good, but it's the total centrality of my friends that matters. My centrality should just be proportional to the sum of my friends' centrality.

So now we just have a system of equations and unknowns. This is just a simple calculation in linear algebra. It's an eigenvector calculation. It turns out that there's nice mathematics behind this. Perron–Frobenius theorem tells us there's a single solution to this that has all non-negative entries, so there's a nice solution to this and you can calculate it. If you go back to this particular network, what do you see? This one comes out at 0.31; this one comes out at 0.11. So this one is sort of three times more central according to this idea that it's your friends who matter, not just how many friends you have.

You can see this one's almost as important as the one that has seven, and the most important node now is the 0.5. So this is picking out different nodes, and this is closely related to page rank and other notions that were important in Google as a search engine and other things. If we think about information spreading, it's keeping track of how well-connected my friends are. That might be a better measure than just counting friends.

What we thought about was, neither of these are actually designed for understanding information spread. Information spreads in a way that has two different features. One, it's somewhat probabilistic. And, secondly, it has a finite time horizon. People get tired of it after a while, it decays, and it stops spreading after some time. So, if you look at the half-life of tweets or other things, sometimes they'll be as short as eighteen hours, sometimes twenty-four hours. There might be a period in which things spread, and then it's old news and it just dies out. And so what we added was sort of a simple measure that has two dimensions to it. We called this *diffusion centrality*.

. . . my centrality is not proportional to how many friends I have, but is proportional to the total centrality of my friends.

The idea behind this is, I look at a particular person. Let's just take a picture. I look at somebody—maybe this is the teacher in the village. I tell the teacher that we're going to come in and spread news about loans. Then this person can spread information, and they spread it randomly. They randomly bump into some friends and here the probability is 0.5. That means that any particular friend they're going to bump into with a chance 0.5 in any given period. So the next day, maybe they bump into half their friends and they tell them about this project.

$T = 4$ means that this is going to happen for four iterations and after that, this process just stops. So if we just do a simulation on this, we could go through with this; the first node told somebody. Second day, they tell somebody else—that's sort of spreading outwards. We run a simulation with these parameters. We get a thirteen for this node. So that gives us some idea of, if this was a random process that had some finite time to it, how far would it spread? And then it dies.

Okay. If you do it, start with another node, you go through this and you run it a series of times. You get a six. Okay, now you can do this for different nodes. There are ways of approximating this calculation and getting a nice calculation from it. What's nice about this measure is it actually nests the other two that we talked about earlier. So when we think about this degree centrality, if I just did this once, right? If I had people spread news once and then it died after one period, it would only reach my immediate friends.

If I had twice as many friends as you did, then I would get twice as many people, on average. So this would be degree centrality. If you run this for $T = \infty$, then it just keeps echoing around the whole system and goes forever. And that turns out to be the calculation behind eigenvector centrality. You can prove a theorem that if this just happens once it's degree centrality. If it happens many times, and P is at least one over the first eigenvalue, then this thing converges to eigenvector centrality.

What we're interested in is, communication often happens somewhere in between. It's not totally viral—it keeps going forever—and it's not as if it just goes one distance one and then people never resend things. And so, now, if we look at the different villages and try to predict what percent of microfinance participation was there, how much can we explain out of the overall microfinance participation across these different villages?

If we just use village characteristics—you know, how big was the village, how many castes did it have, and so forth—you can explain about twenty-five percent of what happened. If you add degree centrality, you get it up a little above thirty. If you use eigenvector centrality, it's slightly more but not significantly more. And if you use diffusion centrality, you can get up to forty-seven percent. That's using a crude version of diffusion centrality. If you actually fit it to the data where you try and estimate that P and a T, it turns out that P and R data are about 0.2. People talked to about one-fifth of their friends in a given time period and T looks like a 3. It goes out about distance three and it just dies. And if you use that, then you can get this up to about sixty-three percent. So it goes even higher if you use that.

Okay. So, what does this tell us? In that paper, which, incidentally, was rejected by an economics journal—

[LAUGHTER]

—we eventually published it in *Science*.[2] So that's not something that's unique to non-economists by any means. In any case, what do we take away from this? The informal network was important in getting the information

2 Banerjee, Abhijit, Arun Chandrasekhar, Esther Duflo, and Matthew O. Jackson, "The Diffusion of Microfinance," *Science*, vol. 341 no. 6144.

out and making sure that things got out in terms of hitting the right nodes was important, but hitting the right nodes in a very specific sense.

Understanding the particular complex system that's at work here, in terms of how information spreads, gives us an idea of who is important in these villages, and using the wrong measures doesn't do very well. You don't explain much at all if you've got a sort of off-the-shelf centrality measure as opposed to one that actually tries to model their specific process that's actually going on among these people in this particular instance.

Okay, that was the first part of the project. And let me just say a little bit about what happened serendipitously in this. We've got a Ken Arrow quote here. So this leads to an important longstanding question: "Does the market (or, for that matter, the large, efficient, bureaucratic state) destroy social links that have positive implications for efficiency?" If you've read Putnam or Bourdieu, you've seen some complaints about the erosion of social capital once we get into large market systems.

Here we actually had a system by which we had sort of a serendipitous experiment where we could really see whether that was true. In 2006, we surveyed the seventy-five villages. From 2007 till '10, they entered forty-three villages, offered loans. They didn't go into the other ones. And in 2011 and '12, we resurveyed all the villages. So then we were able to compare the villages that got loans to the ones that didn't.

Now, this is not a controlled experiment, so we can't do causal inference, because they entered forty-three villages without randomly picking those forty-three villages. Since then, we've also done another study where we went into another 102 villages and we randomly picked the villages that a bank went into and then held a CEP constant. We get more or less the same numbers, but I'll show you the non-randomized version because we have better data here.

So, what actually happened here? First of all, let's just look at the density of the networks in these places. How many connections do typical families have? If we look at the non-microfinance villages, we can look at the fraction of the rest of the population of families that a given family is connected to. So, a little less than ten percent of the families. Afterwards, it was a little more than eight percent.

Okay. There was a drop in those villages, but the drop we saw in the microfinance villages was about twice as much; we saw about a two-times-larger drop in the number of connections in the villages that actually got loans. One story for this is, well, people get loans and now they don't need to borrow from their friends. And so that particular kind of borrowing and lending relationship starts to disappear. The question is, does that disappear more broadly? Are there other networks being affected? So we would expect the borrowing and lending relationships to disappear. And, indeed, they do among the people who got microfinance.

We looked at not only the borrowing and lending relationships; we also broke them down by different types of families. So there are certain families who are likely to take out loans. We call those *high propensity*. We did propensity scoring here to bin these different families. There are some families that are highly likely to get microfinance loans. There are others that are less likely. The highly likely tend to be more educated, a little wealthier, a little older, a little larger, more central. So there are certain aspects that predict whether a family is going to get microfinance.

We can compare those in the microfinance villages to the non-microfinance villages, and then the people who never would have gotten the loans to the ones from the control group as well. What's actually happening is it's not just the high types, the people who are very likely to get microfinance, who are losing these relationships. In fact, the lows are losing relationships at even a faster rate. Moreover, the lows lose relationships not just to highs but to other lows.

Then, when you add in the advice network, the advice network drop looks almost exactly like the borrowing and lending network. In particular, what's happening here is that a lot of their relationships are what are called *multiplex* by sociologists. They're layered on top of each other. The people I borrow and lend with are also the people I get help from, I get my kerosene and rice from, I get my advice from. If we kill that borrowing and lending relationship, we kill a whole series of other relationships. And we're seeing it erode not only among the highs but also among the lows in this population.

So, you know, what are the kinds of lessons we learned from this particular empirical investigation? One is that there are complementarities and spillovers. The microfinance participation behaviors affect others, even the people that they're not directly related to. So when we look at these networks, there can be families that are completely isolated from the microfinance people, and they're losing relationships as well.

We go through modeling and try to understand exactly what's happening here. The story that seems consistent with what the people tell us from the villages is that they sort of hang out in a town square. They go to tea together. There are certain kinds of things that they do together, and when the highs get the loans, they stop doing as much of that. And that sort of erodes that particular interaction, and then the lows stop going there and then the whole thing spirals downward. The town squares in the microfinance villages are just less active and less interactive places than the ones in the non-microfinance villages.

Okay. Summary of implications of exposure to these formal loans: they're changing social fabric and having unintended consequences. The lows are losing relationships and it could increase inequality. We see it in terms of the networks—should we see an increase in the network inequality? We don't have measures here of the financial inequality. We were actually connecting that and some of the other data where we can do a better job of actually seeing whether this also has financial consequences. But, you know, it certainly doesn't mean that we shouldn't be offering loans. It just means that, when we look at any kind of economic system, those economic systems interact with social systems. The social systems are important in making the economic systems function, and the economic systems actually affect the social system structures as well.

That's sort of the message that we pull out of this—these things interact in conjunction. They're intertwined. And when we design policies in one sphere, we should have a good idea of what the implications might be in another sphere. And if we're going to be modeling, as researchers and scientists, we have to think of these things holistically because they're very closely related to each other and they make a big difference. You can read more about this. I have a recent book and sort of an older book that's more technical in nature. Thank you very much.

[APPLAUSE]

AUDIENCE MEMBER Matthew, thanks so much. Isn't what you're seeing happening in these villages just globalization on a very small scale? In other words, the people who are getting the loans—aren't they likely establishing new relationships with people outside of their villages or whoever their customers are, etc., that might not be captured by your networks?

M. O. JACKSON Yeah, certainly. I think there are two aspects. One is part of the fact that, even in the control villages, we're seeing some erosion in the basic social structure. In this area, people are getting jobs in Bangalore and in other places and moving more. More of their life starts to go outside the villages. And we might expect that for the highs—you know, the people who actually got loans. The fact that the lows are losing more of these relationships and these are the people who aren't getting actual access to the credit and not moving out is they're less likely to be moving outside the villages.

I think there's sort of a complex interaction, which says that how one part of your society's structure works depends on the other part. As we change the highs—and it could be that they're forming additional relationships outside the villages—then that's impacting the lows as well. So, yes, certainly. I think on a global scale, I wouldn't want to extrapolate too much from this. It's certainly not a message at all that markets are a bad thing and so forth. It's just that they have impacts and they operate in conjunction with these other systems.

C. MÓNICA CAPRA This is related to the previous question. Do you have a sense whether the villagers felt that they were worse off? Do you have a measure of well-being, for example? You could imagine that even if you are a low type that perhaps you don't want dependence on the mean cousin, you know, who can actually come back and hit you if you don't give something back that you owed. So, in a way, even though the social fabric is being eroded, it could be that it's for the benefit of everyone, although you need more time to see it.

M. O. JACKSON Yeah, I think in long terms we can't be saying too much and we don't have really good welfare measures. We have tracked the borrowing and lending of the lows as well. What does happen is the lowest tend to substitute the types of loans that they would've gotten from friends and family and acquaintances, and instead they're going to moneylenders. The interest rates in these villages for moneylenders tend to be about fifty to fifty-five percent. The interest rates from the bank were between thirty and forty percent. The interest rates among friends and family tend to be between zero and twenty percent, depending on the circumstances. So there are very different interest rates and it does appear, at least crudely, they're moving away from the social structure and you're seeing that loans go up in terms of the moneylenders instead.

There is some substitution that they're doing, but we don't have really good consumption data in these households and so forth. I think the main message is just we see changes and it just means that we should be more aware of these interacting systems, but we don't have a good welfare measure for that at this point.

. . . when we design policies in one sphere, we should have a good idea of what the implications might be in another sphere.

DAVID WOLPERT Hi, Matt. I'd like to focus on what you just said, which you also alluded to at the very end of your talk. This can be summarized as "beware unforeseen side effects." There's a related issue that, actually, the SFI has been trying to grapple with for a long time:

How do you model innovation that's truly innovative, in the sense that you don't build the possibility of that innovation into the model from the first place?

For a striking example of this issue of how-to-deal-with-things-left-out-of-models in economics, we can go back to the great debacle that sullied the reputation of neoliberal economics in 2007. That recession was completely unforeseen, i.e., reflected processes not in anyone's models. Lots and lots of Monday-morning quarterbacking, where we now look back and see all the signals, but at the time it was unforeseen.

Do you have any insights into how to get our hands around this kind of an issue?

M. O. JACKSON I don't have a simple recipe. I think the one thing that we underestimate often, and this has come up already in some of the discussion, is the interaction between various types of systems. Here we're looking at informal social systems and more formal market transactions where you're dealing with a bank, but also with villagers. And that can be true whether we think about the interaction between the government and markets or various political entities or organizations.

So I think that, in general, it just says that any time we're looking at some system, we ought to at least look at the adjacent systems, the ones that we know the participants are also members of and also interacting in. That's sort of a minimal starting point. And when you talk about the financial crisis, you know, the financial system was wonderful for moving capital around and getting it to places it never had been before. But, at the same time, that movement of capital meant that now defaults could cascade. There are all these unintended consequences and I don't have a simple recipe for anticipating these things other than to realize that, anytime we're looking at a system, we shouldn't just do sensitivity analysis within that system. We should do it across systems as well.

TIMOTHY KUBARYCH[3] You used that algorithm that broke it into blues and reds, and there's a reality behind that—it's like you're in one class versus another class. The guy who passed away who used to run Fiat used to say that it was amazing in America, that people would move to places, whereas in Europe people wouldn't. And so it degraded the social contracts that people had with their small communities. But you get all of this—like the room that we're in. So, is there a way of measuring whether or not you're going to leave the group? In that culture, you're not going to leave. You're not going to move from red to blue. But in other systems you will change the group that you're in if the connection between you changes.

M. O. JACKSON Yeah, there's literature that I'm not directly involved in on migration and networks and trying to understand what's push and pull. For instance, if you look at particular countries, you know, people who've looked at Mexico and where people from Mexico go.

3 SFI ACtioN member and Partner, Deputy Director of Research, Harding Loevner LP.

They tend to go to places where they have friends and family already established and where they can move into someplace where they have those connections.

Often when you're actually looking at migration, it could be that I'm cutting some ties in one place, but often it's being pulled from ties in another place that will determine where I end up. That's a fairly strong indicator of what my prospects are. There are really nice studies by Lori Beaman, for instance. There were people who relocated as refugees and she looks at how large the community was of the friends and family and acquaintances of the people who were relocated.

And when they move into an area where they have an existing network, they had significantly higher employment rates and integration rates than in situations where they were moving into some place where they didn't have any connections. Those kinds of informal social structures are very important, but they tend to be really highly segregated. They tend to fall on all kinds of lines. Here we're seeing it along caste, which is obviously very strong in India. In the US we see it along ethnic lines, along religious lines, along political lines, increasingly along age lines. It's pretty hard to find a human network that doesn't split along some kind of divide and those divides are really important in determining where people end up and where they go.

[APPLAUSE]

8

COMMUNICATION & COORDINATION IN EXPERIMENTS

Introduction & Talk by C. Mónica Capra

There is no doubt that the internet has dramatically changed the way in which we communicate and interact with each other. Social-network platforms such as Twitter and Facebook allow us to share information with a large group of people at almost no cost. A simple entry on a Facebook timeline can go viral as friends share the entry with their friends, and the friends of friends share with other friends. But what has been the effect of these novel ways of communicating on human actions? On an intuitive level, new communication technologies improve our ability to resolve collective-action problems, which happen when a group of people want to take an action together to achieve a common objective. Protests are an example of collective action. People participate in protests if and only if enough other people join them. An individual protesting alone risks ridicule or prison, and little is gained from just one person protesting. In contrast, thousands or even merely dozens of people protesting reduces the risk for each individual and increases the chance that the movement will succeed. Indeed, new communication technologies seem to facilitate coordination.

To see the effect of communication platforms on coordination, my co-authors and I take an approach that begins with a simple theoretical model and ends with a complex-system simulation. The theoretical model is an abstraction, or, simply put, a caricature of real-life social networks where different kinds of communication technologies can exist. Starting with the theoretical model as a foundation, we design laboratory experiments with human subjects. In the experiments, human

---

OPPOSITE *Economic Curiosity: Francis Crick's Letter to His Son*

subjects make decisions under different network structures and communication technologies. The objective of the experiments is to see how different ways of communicating affect real people's willingness to participate in a collective-action task. Observation of human actions under different experimental conditions helps us identify behavioral patterns. We then embed the identified behavioral patterns into algorithms within an agent-based model (ABM). The ABM helps us uncover what would happen (e.g., the location within a network where protests would appear) if agents followed the identified behavioral rules in complex, real-life social network interactions.

In the 2019 SFI Symposium presentation that follows, I explain a simple theoretical model of communication based on Chwe (1999) where players within a network activate if they believe a threshold number of other players are also active. This is a workhorse model for our experiments. The model identifies common knowledge (a string of embedded levels of knowledge) as a necessary condition for successful coordination and collective action within networks. I also describe three laboratory experiments. In the first one, human subjects interact with four programmed bots. The second and third experiments consist of groups of five human subjects interacting with each other to resolve a collective-action problem. In the three experiments, my co-authors and I test whether communication (how one exchanges messages and what the messages are) influences outcomes. I do not cover the results of the ABM simulations in the presentation, but interested readers can refer to our work in Korkmaz et al. (2018). The second experiment is an ongoing project and, once all behavioral data are collected, we will use ABM to see how agents would behave in more complex, real-life environments.

There are three broad takeaways of my presentation. First, we study complex social interactions, like protests, with a systems approach. More specifically, we start from an abstraction or caricature model to identify mechanisms, test these in a controlled environment to identify behaviors, and embed identified behavioral rules into ABM to add complexity. Second, common knowledge (what one knows about what others know) matters for collective action within networks. Common knowledge has been all but ignored in the literature. For example, common knowledge is not present in social movement diffusion models (e.g., Centola and Macy, 2007). The third takeaway is relevant in the context of current events. Our study suggests that Facebook, Twitter, WhatsApp, etc., are likely to make protests occur more frequently and spontaneously. Through these communication technologies, both real and imagined grievances become common knowledge within diverse sets of social-network architectures. Common knowledge, in turn, creates the necessary conditions for collective action.

During the Q&A session, someone commented on the possibility of using simulated environments and evolving AI agents to see if certain kinds of messages are more likely to result in collective action. Messages that lead to protests, for example, could be very emotionally charged. I believe this is an interesting idea to pursue. One could identify network architectures and communication structures within social networks that are ripe for common knowledge. Infecting or seeding these networks with emotional grievance narratives would likely generate protests.

Further Reading

Capra, C. M. and T. Tanaka. 2007. "Communication and the Extraction of Natural Renewable Resources with Threshold Externalities." In Cherry, T. L., S. Kroll, and J. Shogren, eds. *Environmental Economics, Experimental Methods.* Routledge.

Capra, C. M., T. Tanaka, C. F. Camerer, L. Feiler, V. Sovero, and C. N. Noussair. 2009. "The Impact of Simple Institutions in Experimental Economies with Poverty Traps." *The Economic Journal* 119 (539): 977–1009.

Korkmaz, G., C. M. Capra, A. Kraig, K. Lakkaraju, C. J. Kuhlman, and F. Vega-Redondo. 2018. "Coordination and Common Knowledge on Communication Networks." In *Proceedings of the 17th International Conference on Autonomous Agents and MultiAgent Systems*, pp. 1062–1070. International Foundation for Autonomous Agents and Multiagent Systems.

Korkmaz, Gizem, Teja Pristavec, C. M. Capra, V. Lancaser, C. Kuhlman, and F. Vega-Redondo. "An Experimental Study of Common Knowledge and Coordination on Social Networks." Working paper available upon request.

WILLIAM TRACY Early in the talk that Eric gave us at the beginning of the day, we all sort of had a good chuckle at Francis Edgeworth sipping brandy and pontificating about how economic agents actually behave. If we think of that as maybe the least rigorous side of the spectrum for determining agent behavior, our final speaker today works at the most rigorous side of the spectrum. She does work at the intersection of experimental economics, behavioral economics, and neuroeconomics. She's a professor at Claremont Graduate University. To give our final talk for the afternoon on communication and coordination and experiments, C. Mónica Capra.

C. MÓNICA CAPRA Thank you so much. So, I'll talk about communication and coordination in experiments and, just to give you an overview of what our research strategy is, it kind of mirrors what Will was talking about this morning. You come up with an abstraction, which is an economic model of behavior, then we take this economic model of behavior as a way to design an experiment with human subjects. From the behaviors of human subjects, we can derive the system rules that later we give to computer scientists to code into agents in an agent-based model.

So, the agents can behave in more complex environments with the decision rules that were extracted from human subjects. In terms of communication, communication is super important to human societies, as we all know. It's important for exchanging information, it's important for human bonding and culture, but another important role that communication has is in coordinating actions. An example of coordinating actions is protests. So, protests, as David mentioned earlier, are going on all over the world. Hong Kong is the most notorious one. Protests are what economists call collective-action problems, where an individual doesn't benefit from taking an action unless enough other individuals take the same action.

One individual protesting, let's say in Hong Kong, not only risks imprisonment, perhaps even death, and the likelihood of achieving whatever goal he wanted to achieve is pretty low. Now you can imagine a thousand, hundreds of thousands, or millions of people protesting, then the risks for the individual of going out to protest are dissipated.

And the potential benefits, in terms of perhaps achieving the goal of regime change, whatever it be, are enhanced.

So, we can conceptualize this as a kind of coordination game with many different potential outcomes, and some outcomes are more desired by the people involved than others. In the case of Hong Kong, perhaps, everyone wants to go out and fight for freedom. But if people believe that not enough people will be out, then it's crazy to do so. Right? So people would choose to stay home if they believe the others are going to stay home, and will choose to go if they believe the others are going to go.

So, what's required in order for community to have this kind of collective-action solution, or coordinating the better outcome for those involved, is some type of institution or mechanism that will get them to do one thing. Obviously, communication is the one we can think about in human societies to come up. But what's interesting and not so obvious is that not all communication actually leads to effective coordination. So it matters how communication flows within a group of people.

One individual protesting . . . risks imprisonment, perhaps even death, and the likelihood of achieving whatever goal he wanted to achieve is pretty low. Now you can imagine a thousand, hundreds of thousands, or millions of people protesting, then the risks for the individual of going out to protest are dissipated.

The reason is, in order for communication to be effective, it has to create what's known as common knowledge, which is a string of embedded knowledge—you know, if you have two individuals, and I need to know that the other is going to show up to protest, the other needs to know that I'm going to show up to protest, and I need to know that he knows that I know, and so on. So, communication does not always generate that required common knowledge.

So, this is an example from Chwe.[1] Imagine we have a neural network, you know, one, two, three individuals. And let's say there is underlying grievance here. They all want to protest. Now, they know their own threshold. Threshold refers to how many additional individuals need to show up in order for them to show up. If the threshold is not met, it's better to stay home. If the threshold is met, then you want to go and protest. For simplicity, we are going to say that all these individuals have a threshold of two. They need two people—so, an additional individual—to protest. They know their own threshold, but they do not know the threshold of the other individuals. In the first case we can say, suppose One sends an email to Two and Three. They all know that there is some grievance going on; they just do not know the state of the world.

So, One sends an email to Two and Three. Two replies to all. Three doesn't respond. The question is, who is going to protest? We can see that the answer to that question will depend on whether common knowledge is generated. One and Two will protest because One knows that Two knows that One is going to; it has the threshold met. And Two knows that One knows the same. Three, on the other hand, did not respond to the email. And we all have those colleagues who never respond to emails, right?

[LAUGHTER]

So, will Three protest or not? The answer is yes. And the reason is because he knows that Two and One have a threshold of two because he got the messages and he knows that One and Two are connected. So, he can predict that One is going to participate. Likewise, he can predict Two is going to participate. Therefore, he will participate.

In the second case, on the other hand, Two doesn't reply to all. Two just replies to One. Will Three participate? The answer is no. And the reason is because, as far as he is concerned, Two could have a threshold

1 Chwe, M. 2000. "Communication and Coordination in Social Networks." *Review of Economic Studies* 67 (1): 1–16.

of four or five. He doesn't know actually that One would participate because it could be that Two has a threshold higher than two.

He doesn't know the same level of information he has in the previous case. Now, in the other case—in the last case we have that One and Two send a message and leave, let's say, a voice message. In this case, Three knows the thresholds of One and Two, but there is no common knowledge, so Three will not participate. Here we have a kind of distinction between knowing what your neighbors are going to do versus creating common knowledge with your neighbors. The bottom line is that small changes in the communication structure can affect decisions.

Let's think of other communication structures. The previous one can be sort of bilateral communication, like email, but we can have kind of more complex types of communication. Like Facebook, where you have timelines and what can happen is, your friends can post messages on your wall or your timeline and they can actually read what your friends are posting. So, in this example, Red had posted on his wall and Green, even if Green and Blue are not connected, can actually see what their thresholds, let's say, are through the wall of Red.

The bottom line is that small changes in the communication structure can affect decisions.

So, what the kind of Facebook wall communication is doing is allowing these friends of friends to actually have the connection. So, not only Hong Kong is having all these protests, but we are seeing protests going on all over the world, especially the Spanish-speaking world. For some reason it has a tendency to protest. But we have protests going on in Spain, we have protests going on in Chile, in my country, Bolivia, in Lebanon, Iraq, and so on and so forth.

You may wonder, what's going on? Are people going crazy? What's likely to be happening is that there is clearly a sense of grievance among individuals and there has always been, but what's happening is that the kind of communication technologies that we have today, like WhatsApp and so on, are generating this sense of common knowledge of grievance.

And people are realizing, "Oh my gosh, it's not only I'm depressed but all these people are depressed and we need to do something about it."

In addition—this is not something that we've studied, but we'd like to study—you could generate narratives through these communication platforms, narratives that become your reality. They're more often than not realities of grievance. The bottom line is the knowledge about what people know about what other people know is crucial for taking action. And the way people communicate is important for generating common knowledge within a group of individuals.

So what we do is we take this simple kind of model of common knowledge to the lab. We have done two experiments. We recruit people from Amazon Mechanical Turk, and in the first experiment we group people in groups of five, but four of the five are actually program bots. So we're only observing the behavior of one individual.

The program bots are programmed so that they will always play what's optimal to them. We look at two kinds of communication, no communication and bilateral communication, and then we also look at knowledge of the network. The individuals can have local or global knowledge. In the local network, you just know who your friends are. There are some previous studies that show that people do not need to know the global structure. They behave the same when they have local information as well, and global information may not be needed.

But in the second experiment, we recruit subjects and we have groups of five people playing. They are all human subjects and the choice is whether to participate or not in all of this. If they participate, when the threshold is met, they make more money. If they don't participate, they make $75 experimental or $50 experimental. But if they participate but the threshold is not met, they make zero. So, participation is risky in that sense.

Now, in the all-human-subject experiment, we did three types of communication: the control, which is no communication; wall communication, which is Facebook; and then bilateral communication. The communication structure and whether people knew the local or global, that's kind of the main distinction between subject design that we have. But we also have, within one session, the same individual making decisions in different kinds of networks: a star, a circle, and a clique.

Now, you can see that the interactions between how people communicate and the network they're in affects behavior. For example, if you're in the star network with wall communication, you should be expected to have the same kind of decisions as, for instance, clique and bilateral communication. So we're looking at these interactions between the way in which people are kind of talking to each other and the network they're operating. For these experiments, we have preset messages. They are very simple: they're signaling intention to participate.

In each round of the first experiment people were assigned an avatar and a threshold, either high or low. They were assigned one of the kind of networks. And then, depending on which condition they were in, they could send a message or not. In the second experiment with all human subjects, it was pretty much the same except that, for the wall communication, what can happen is you could click on your friend's avatar and you would observe what other people have posted on their wall.

So, these are some results for the first experiment. Overall, what we see is that people with high threshold tended to over-participate. By participating too much, even when they shouldn't, they actually lost a lot of money, or they didn't make as much money as they could. If you pay attention, for instance, to this bar, here is the percentage of subjects who choose to participate—just kind of the red bar. So, this is the case for starting it with high thresholds and the global noncommunication when this individual had all high labels.

We have in the star network five people with high threshold. They shouldn't participate, but there was a high level of participation. When we added communication, and the first experiment is bilateral communication, we see that the participation rates go down. Another thing that we observed is that, without communication, we see that in general there seems to be little difference between the local and global conditions. In general we can say that what bilateral communication did in our experiment is kind of improve the decisions of individuals. Before the communication, it seems that there was an under-participation by the low types and an over-participation by the high types. But after communication, it went closer to the theoretical prediction.

Now, for the second experiment, we are still running the experiments, but we have preliminary results that both the structure of the network and the messaging condition influence the decision to participate. So, what next? We started with an econ model, which is an abstraction.

An individual is very complex, but you have a cartoon that just captures the basic kind of essence of the individual. And that's kind of what the economic model does. But, more importantly, it's allowing us to see the mechanisms. So common knowledge is the mechanism whereby these individuals are actually coordinating their actions to participate. And in the experiments we can derive behavioral rules and heterogeneity as well.

Individuals not only over-participate or not. There are some who are pretty good at making decisions. We actually did use some behavioral rules to add into an agent-based model. In the agent-based model, basically you have agents transition from non-active to active. In the theoretical model, as long as they're given a threshold and a network, you could predict if there is a common knowledge created, that the individuals will transition from, let's say, inactive to active.

But when we have the behaviorally enhanced model, that transition is given by the probabilities derived from what we observe in the experiment. So, the other thing that we would like to do as a research strategy is to kind of do model enrichment and push the envelope. The main thing about this is that the messages were preset. They were just intentions to participate or not.

So, what if we have unrestricted free communication? In this case, it's a kind of a different decision environment. This is a macro economy, actually. It's based on a growth model of the economy. So it's a growth model with a threshold externality. We have here in an economy that if capital is accumulated to some degree, you can get to a better equilibrium than another. So we have an economy with two equilibria. One is an optimal steady state and one a poverty trap. We get subjects to participate in an economy where they have to decide how much to save and consume. And if they consume in one period they get money. But if they save, that saving could become capital for the next period.

If you never saved, you can actually reach the threshold and go to the better kind of equilibrium. This is an economy that has many periods. But every time there is a twenty percent probability that the economy is going to end. So, there is this kind of a risk. If you don't consume and save, you actually could be losing because the economy could end. This is kind of the timing of decisions, but the main thing to look at is, five individuals are participating in this economy, and basically the decision is how much to save, how much to consume.

> An individual is very complex, but you have a cartoon that just captures the basic kind of essence of the individual. And that's kind of what the economic model does.

You could think of them being in a kind of complete network. And the type of communication they have is that they are posting messages on their board for everyone to see. Let me show you the result. The baseline treatment is basically the economy without institutions that will help in coordinating actions. And the treatment we have is institutions, which is communication, exchange of information. In this example, this dashed line represents whether the economy was able to reach the threshold so that it will go to the better equilibrium.

In this economy—despite the fact that we have people talking to each other and everyone can see the messages, so we have that kind of richness and kind of communication—we see that none of the individuals except one was sending relevant messages. The typical message that people were using was things like, do you think they will give us bagels or donuts? I am hungry.

[LAUGHTER]

The fact that there is that communication structure doesn't mean that people are actually going to be talking about the issues.

In the cases where people succeed in kind of coordinating their consumption decisions so they don't consume too much and save enough so they can get out of the poverty trap, what happens is we have more active messages—messages that are more related to the willingness to coordinate—that are relevant. What's interesting is that you can see this number represent how many people are just chatting about the relevant issues, that after a while it's not necessary to talk anymore. And here you have individuals maybe talking about bagels and cookies or whatever. It doesn't matter anymore because you have generated that belief that everyone was going to save enough.

So, what are the patterns of successful communication? We have this specific pattern—and this is just one example, but we have other sessions like it—someone suggests not to consume, and others express their agreement. So, awareness of the problem, a common knowledge, awareness that others know, you know that the others are aware of the problem, the generation of common knowledge. But also what's interesting, we observe that it seems to be that this type of message is, "Good job, well done, we are a team, that's the way to do it." Those kinds of messages tend to help in succeeding, in coordinating, in the better outcome.

. . . what people say, how they say it, and the language they use helps them develop a narrative—it's that narrative that seems to be important for successful coordination.

What we observe is that, although we have this very connected environment where people can send messages to each other, they don't necessarily use it in order to coordinate and create common knowledge. But even when there is common knowledge, in the sense that it seems that everyone knows what the problem is, it's a necessary but not sufficient condition. What you need is some type of an emotional response to what's going on.

In general, what we can learn from these studies is that things are a little bit more complex than we think. That when we are thinking about communication—what people say, how they say it, and the language they use helps them develop a narrative—it's that narrative that seems to be important for successful coordination. Thank you.

[APPLAUSE]

AUDIENCE MEMBER I wanted to ask, where is the threshold? It seems like coordination started happening when people started believing in a team or something bigger, or it was a little bit easier. What's the threshold when people started migrating from just self-interest to, say, like, the greater good, or whatever it is they do.

C. M. CAPRA In the last experiment, the teams just were not talking about saving or consuming at all, as if there were complete unawareness of the problem. So that was kind of interesting, but even when you had some people saying, "Hey guys, we have to make sure that we save enough so that we get better," a lot of times that didn't get any response.

You just have to have kind of the right team, the right people. And here the individuals are randomly assigned to the groups. An interesting follow-up question will be, what if you have a pre-selection of the teams that were more likely to generate this kind of solution to the coordination problem versus just not doing anything about it.

AUDIENCE MEMBER In one of your studies, you had stars and cliques. Obviously, the structure that is used for communication makes a big difference to whether people reach the threshold or not, but actually in practice, could you expand on that a little bit? Because you kind of covered it, but you didn't go into depth. So could you say a little bit more about that?

C. M. CAPRA Sure. If you are in a star network, your friends, let's say, are not connected to each other. Imagine that all of you have a high threshold. Basically, high threshold is threshold of three, which meant that you needed three other people in order for you to actually make it worth it to participate.

Nobody should expect that the other would participate, right? Because they don't know. If you are in a clique, on the other hand, you still have the five people, but they're all connected. So, in that sense, that connection is what allows them to kind of learn about each other's thresholds.

AUDIENCE MEMBER So, in practice, the implication of that then is that something like Facebook transforms the likelihood of protest? I mean, is that the conclusion you're reaching?

C. M. CAPRA: Yeah. Okay. So, the conclusion is you can basically come up with what's called uncommon knowledge sets. What are the substructures that generate common knowledge? If you have a bilateral communication, it turns out that in order to have common knowledge, you have to have sets that are kind of like the clique. Under threshold—in this case, the maximum threshold should be five, right?

But if you have situations of wall communication, like in Facebook, it turns out that the structure that you need, and this is kind of like the theory that plays a role is actually a biclique. So the star will generate common knowledge if you have this wall communication. And there are several papers on the kind of theory behind that. So why is it relevant? It's because if you have a real network you cannot identify the bicliques, let's say.

And in a Facebook type of communication, you know, bilateral type of communication, you can identify the cliques and see that those are actually common knowledge sets. Now the dynamics is another thing that's also unsure.

AUDIENCE MEMBER In your studies, did you actually figure out how small a group of committed individuals could actually precipitate a gigantic protest?

C. M. CAPRA Oh, not really, but this is interesting. This common knowledge kind of approach to looking at how people take collective action. You know, you can have these pockets of common knowledge across different parts in the network. So, you can have, let's say, participation everywhere, right? You don't need, kind of, any seeding, in a way. You don't have to have seeding in order for these things to happen. But, to answer your question, no. I haven't done that.

AUDIENCE MEMBER Thank you.

BLAKE LEBARON In your communications experiments at the end, I was very interested in . . . Would it be possible to eventually put in sort of an AI bot in there that's kind of trained up on the statements they have, but then you give it a kind of predisposition to savings and see if it would push them?

C. M. CAPRA Yeah, that's an excellent idea. Right. It will be interesting to kind of get people to see how is it that—

B. LEBARON Load it with an influencer.

C. M. CAPRA Yeah. And what kind of narrative you could use, you know, like a pattern of messages that will be sent that actually will help them. You know, those people who are like, "bagel, I'm hungry," kind of awaken to the problem.

AUDIENCE MEMBER I was wondering also about ability to analyze the narrative structures of the communications that was most effective. You probably need much bigger samples and so on.

C. M. CAPRA Exactly.

AUDIENCE MEMBER But, just from eyeballing the very small sample you showed, it was interesting that there is sort of commonality of goals, of "let's save together," or even . . . interesting. The comment about bagels, it sounds frivolous, but it could be interpreted as a "We have something in common. We're both hungry so you can trust me," and so on.

Now, one data source you might want to look at that might be of interest as a natural experiment. When traders were illegally coordinating to manipulate LIBOR, the interest rate in the UK, the regulators released a huge trove of transcripts of their private messages over, I think the Bloomberg private messaging network. And it looked a lot like that actually *[laughs]*.

C. M. CAPRA Oh, really?

AUDIENCE MEMBER Except instead of saying "We want bagels," they were all talking about the champagne they would share after they manipulated the rates.

[LAUGHTER]

But it was interesting. The narrative structures struck me as similar.

C. M. CAPRA Yes. I guess it's the point that they've made in the beginning that it's kind of weird to see individuals as if they were just agents.

We know complexity, and I think we individuals are kind of complex. Like, we will not—let's say, I don't think—behave. We have the psychology and the need to build stories, right?

In terms of communication and collective action, it's super interesting to figure out what it is that people are making, what story is getting them to take an action. I don't know if people are studying this thing—oh, you are? Okay.

AUDIENCE MEMBER I was wondering if the two examples that you gave—in one case the collective action is pursuing a positive goal. And in the protests it's responding to a perceived threat. And in the political violence literature, it's using the perceived threats that are much more effective in spurring collective action. So, I was wondering if you see anything structurally different, or would you expect to see something structurally different in some of the thresholds or the types of structures? If you're really differentiating and controlling for whether it's a positive goal or a response to a threat, you know, avoiding a negative consequence, is one question.

And the second is, do you have plans to really try to pursue this? I think that that distinction between the information content—just in terms of information versus the emotional content and what that might mean—is we have mixed systems of people and AI and bots as communication networks, and presumably the AI and the bots are not going to be responding as much emotionally, but the people are, and what that might look like and how that might make things different.

C. M. CAPRA Yeah, I would agree with you that, given that we respond differently with respect to the kind of gains and losses, I expect we'll see a difference if there is more of avoidance of a threat other than achieving more money or a better benefit. I would think that people would be more willing to take the risk if there is threat. Right?

In terms of the second question, I agree with you. We did the analysis, the kind of the coding of the emotional content of these messages, and even if common knowledge was there, basically everyone was talking about saving more. Everyone agreed that we should save more. There was not this emotional kind of bonding or, you know, "Well done, we are a team," stuff like that. Communication was necessary but not sufficient.

Okay? So, you needed this, and that was kind of a bit surprising to us. Maybe this is an obvious thing. But yeah, with bots participating, I guess bots could also express emotions in a way.

Okay. Thank you.

[APPLAUSE]

2

COMPLEXITY ECONOMICS: WHY DOES ECONOMICS NEED THIS DIFFERENT APPROACH?

Introduction & Talk by W. Brian Arthur

This talk was the keynote for SFI's 2019 Applied Complexity Network Symposium, held post-cocktails on the evening of the first day. Complexity economics was very much birthed at the Santa Fe Institute in the late 1980s and the 1990s, and indeed for several years it was known as Santa Fe economics. This talk is an informal attempt to tell how complexity economics got its start at SFI, and to make sense of the logic of its approach. I have left out many things about modeling and people and early agent-based work in our program, and our early probability approaches. More details are given in other accounts.[1] But I want to emphasize here —which I didn't in the talk—that SFI's ideas arose in a context. They had precursors in the work of Peter Allen and Ilya Prigogine of the Brussels group, nonlinear dynamics, nonlinear stochastic processes, and in theories of induction. The Ann Arbor group directly influenced us through John Holland.

I'm struck that so much has happened since these early days at SFI, both within SFI and outside. Rob Axtell and Josh Epstein introduced their "generative social science." Network theory took off. Agent-based

1 For example Waldrop, M. 1992. *Complexity.* (Simon & Schuster, NY); Fontana, M. 2010. "The Santa Fe Perspective on Economics," *Hist. Econ. Ideas*, 18, 2, 167–96; Arthur, W. B., 2010. "Complexity, the Santa Fe Approach, and Nonequilibrium Economics," *Hist. Econ. Ideas*, 18, 2, 149–66; Arthur, W. B. 2015. Preface to *Complexity and the Economy.* (Oxford Univ. Press, NY).

OPPOSITE *Economic Curiosity: Tulip Mania*

computational economics[2] took off. Models that use machine learning and AI to simulate agent behavior have shown up. My own interests turned to theorizing about how technology comes to be, influencing how economies form and re-structure themselves.[3] Some of these new threads were reflected at the symposium, some were not.

Whatever was included in these symposium talks, complexity economics has now become a large field. And the Santa Fe Institute—particularly Kenneth Arrow, Philip Anderson, and David Pines—deserves much credit for early on backing this work. I am surprised that our small program, "The Economy as an Evolving Complex System," has come to have so much influence. It was SFI's first research program, and we were aware—at best vaguely aware—that we stood some chance of doing economics differently. I also think the times deserve much credit. Economics changes with change in technique. And just as algebra and calculus entered economics in the 1870s and brought in neoclassical economics, in the 1980s we all got desktop computers and that helped birth this new approach. We could suddenly allow ourselves to get interested in more complicated models—computation could handle complication.

In fact, since this talk was given I am struck that computation has gone from being slightly outré to being now highly fashionable. There's been an enormous burgeoning of agent-based computational economics (ACE) models in the last few years. This is healthy, I believe. In good hands, the sort of detail computation allows makes for more realistic modeling, and we are seeing this currently in coronavirus models. The old equation-based models with buckets of Susceptibles, Infecteds, and Recovereds are too coarse, and computation allows detail on how precisely infection spreads and how it's related to the niceties of economic arrangements. So we are learning a lot from models that allow more detail.

Does this mean the complexity-economics approach will be eclipsed—shut out—by all this enthusiasm for agent-based modeling?

---

2 See Tesfatsion, L. and K. L. Judd, eds. 2006. *Handbook of Computational Economics: Vol. 2. Agent-Based Computational Economics*, North-Holland Elsevier, NY. Hommes, C. & LeBaron, B., eds. 2018. *Handbook of Computational Economics, Vol IV: Heterogeneous Agent Modeling*. (North-Holland, Amsterdam).

3 Arthur, W. B. 2009. *The Nature of Technology: What it Is and How it Evolves*, Free Press, New York.

I don't think it will. There is a symbiosis between the two. Complexity economics heavily uses computation, and in particular, agent-based modeling. And ACE, when it needs a theoretical foundation, appeals to complexity economics. ACE is a method, or set of methods, and as such it is not an approach to economics any more than algebra is. So, complexity economics, I believe, in whatever form it develops from now, is here to stay.

The reader may recognize that much of the material I talk about in the following I have written about elsewhere. And there is some overlap with the introduction chapter to this volume. I have tidied the talk a bit but kept the roughness of it to some degree—it's a talk, after all.

DAVID C. KRAKAUER Brian is the original maverick. He left his position at Stanford, where he was a highly decorated professor. He'd won the Schumpeter Prize and the Lagrange Prize and he came here to remake economics. He's an inspiration for a lot of us, including many of you here. Brian, very much looking forward to your talk.

W. BRIAN ARTHUR Good evening. What I want to talk about tonight is how complexity economics came to be, at the Santa Fe Institute here, what it's about, and above all what difference this new approach makes—if any. People often say to me, "You have this complexity economics or nonequilibrium economics—call it what you might—but isn't it something of an add-on to economics? It gives agent-based models, and emergent phenomena, and self-organization, and all that, but isn't it really a bolt-on to standard economics? So, what does complexity economics provide that's different, that would change the way we see or understand the economy or change policy? Where does it fit in? And how should we think of it?"

Let me begin from some basics. The economy, whether in Europe or the US, is an enormous collection of arrangements and institutions and technologies and human actions, buying and selling and investing and exploring and strategizing. It's a huge hive of activity, where the individual behavior of agents—banks, consumers, producers, government departments—leads to aggregate outcomes. A couple of hundred years ago, economists such as Adam Smith noticed in this that there was a loop. Individual behavior leads to aggregate outcomes, and aggregate outcomes—the aggregate patterns in the economy—cause this individual behavior to adapt and change. The individual elements of the economy in other words react to the patterns they cause.

It is this loop that tells us the economy is a complex system. In this sense, complexity is not new in economics. As in all complexity, individual elements (human agents in this case) react to the pattern they create. Complexity, in a phrase, is about systems responding to the context they create.

For the first hundred or so years of the field, many economists thought in these terms, but this view wasn't easy to formally analyze. So economics came up with a simpler way to look at things. Starting about 1870, it asked what individual behavior would lead to a collective outcome that would validate or be *consistent with* that behavior—what outcome would give behavior no incentive to change? The situation would then be in stasis or *equilibrium* and we could much more easily look at it via equations and mathematize economics. We could keep the object still while we were examining it. This was a kind of finesse; it was clever because if we can simplify the economy into an equilibrium system, we can reduce economics to algebraic logic. If you are willing to assume, in the part of the economy you are looking at, that individual behavior produces an outcome that doesn't give that behavior any incentive to change, you have the idea of a solution. This worked. It's a brilliant strategy: it gives insight, you can teach it, and you can use elegant mathematics. I was trained as a mathematician and this attracted me into economics. I thought I could make a killing because I knew quite sophisticated mathematics.

It's worth looking more formally at how this equilibrium finesse works in modern economics. You take some situation in economics, whatever it might be; it could be the theory of asset pricing, or of insurance markets, or international trade. You define this as a logical problem. You assume each category of agents—they might be consumers, or investors, or firms—has identical agents. They are all the same, and they know that everyone else is identical to them, and they all face the same problem. Further, you assume the problem is well-defined, and agents are infinitely rational, so they can optimize given the constraints of the problem and arrive at the best outcome for themselves, knowing others are behaving the same way. What overall outcome validates and gives no incentive to change their individual behavior? This is the basic recipe for doing theoretical economics, and its modern version is largely due to Paul Samuelson and a few economists who preceded him. The recipe is quite sensible in a way. It gives us a logical approach that can be made mathematical, and therefore scientific.

THE STANDARD APPROACH

1. Well-define a situation as a logical problem.
2. Assume identical agents who behave identically, rationally, and who know that other agents are identical and behave identically and rationally.
3. Assume agents' behavior produces an aggregate outcome that validates their individual behavior.
4. The problem becomes mathematical.
5. The solution is an equilibrium.

Slide 1

I've put this in a slide (slide 1) that's a bit complicated. The payoff is that if you can understand this slide it's worth five years of graduate school in a good economics department, because this is the standard method they teach you.

And so everything in economics is brought into this framework and put through this form of analysis if it's to be seen as theory. If agents were identical drivers of cars in crowded traffic, this is like asking what speed would produce a traffic flow where no car has any incentive to change. It has a rather lovely property of everyone achieving what is best reachable for them in an outcome where everyone else is attempting to achieve the best reachable to them. Any time I look at this, I find it beautiful. It's elegant. It's pure and it's perfect. You can prove theorems showing that if people behave this way, it'll be in some way optimal or efficient and it's highly mathematical. I want to say that this particular way of looking at the economy—it may be mechanistic; a lot of it was borrowed in the late 1800s from physics—has been useful. It has allowed economists to solve problems of finance, of central banking, of industrial production. We understand markets, we understand international trade, we understand currency regulation, and many other issues. We really understand large parts of the economy and we're not having major depressions like we had in the 1930s. This way of looking at the economy has achieved a lot.

But of course there's a "but." A very large "but." And that is, if we assume equilibrium so that we can do mathematics, it puts a very, very strong filter on what we can see. If we can only look at equilibrium situations there's no scope for adjustment or for exploration or for creating startups or for any phenomenon in the economy, like bubbles and crashes that appear and disappear like clouds in the sky. There's no structural change; there's no unexpected innovation—they're not equilibrium things. And, if everything is the same over time, time disappears and there's no history, so there's no path dependence. We accepted equilibrium because it is so analytically useful, but it gives us a Platonic universe. It's beautiful, and ideal, and pristine, and lovely, but it's not really real. So, economists in the last fifty years have begun to question equilibrium. They're questioning the standard theory's authenticity. Once, in the early days at Santa Fe, I asked the computer scientist John Holland what he thought of equilibrium. Anything at equilibrium, said John, is dead.

So the question is, is the economy more alive, more vital, than equilibrium would show us? This was the question that landed in the late 1980s in Santa Fe.

In 1987, the Santa Fe Institute was in the old convent on Canyon Road and it was very much a startup. My colleague Kenneth Arrow at Stanford brought me there, and he brought several other economists there. He and the physicists David Pines and Phil Anderson had brought some scientists to meet their counterparts in economics. One of the people on the scientific side was John Holland. Another was Stu Kauffman. There were quite a few people: mathematician David Ruelle and physicists Richard Palmer and Doyne Farmer. On the economic side we had Larry Summers, Tom Sargent, Buz Brock, and other eminent economists.

The two groups met and paraded ideas. This went on for ten days. It was awkward at first but in the end extraordinarily successful. We got a kind of sugar high on the ideas, and the Institute, after this now-famous conference, decided it would start its first program that was going to be funded by Citibank, by John Reed. Arrow and Anderson and Pines would be the godfathers of this, and they asked Holland and myself to run the program. John couldn't get away from Ann Arbor, so I was brought in on a sabbatical from Stanford to lead this program. That was the good news. You know, "Brian, you're running a program on 'The Economy as a Complex System.' You're going to have the equivalent of a couple of million dollars, which allows you to buy out quite a few people on sabbaticals. You can have Arrow and Anderson (both Nobel Prize winners) do the inviting so nobody is going to say no. You can bring them all to a convent in the Rockies. They can all think what thoughts they want and nobody's going to object. It'll be isolated and nobody's going to know what you're doing."

. . . if we assume equilibrium so that we can do mathematics, it puts a very, very strong filter on what we can see.

The bad news was that when we got together in the fall of 1988 we weren't sure what to do. I remember Ken Arrow saying, "Well, we could do something in chaos theory. That's interesting." I was thinking we could do something on network effects or increasing returns, but I'd already spent many years doing that and was tired of it. We thought maybe we could take spin glasses or something from physics and translate that into economic terms. Nothing was quite settled, and this went on for about a month. We'd meet in the kitchen of the old convent. We'd have coffee, and we'd say, "Okay, what's the theme going to be? What's the direction going to be?" Finally I called Ken in Stanford to ask him what direction we should move in, and he called Phil Anderson, who called our funder, John Reed, the chairman of Citibank. The word came back, and it was unexpected. "Do whatever you want, providing it is at the foundation of economics and is not conventional." I was stunned. I couldn't believe our good fortune.

So we had permission to do what we wanted but still didn't know what that might be. Strangely, it was someone who didn't know anything about economics, Stuart Kauffman, who broke the logjam. One morning Stu said, "You know, I'm listening to all of this," he said, "and you keep talking about equilibrium. It's like a spider's web where everything's in equilibrium, the economy, but what would it be like to do economics *not* in equilibrium?" I tried to quiet Stuart. I thought this wasn't a good idea. This was freaky. We weren't going to mess with that. It was a kind of third rail in economics and I didn't see how we could do it. But Stu had planted a seed. It started to grow in my brain and in other people's brains and we started thinking seriously about it.

Let me show you for a moment what it's like to think outside equilibrium. Here's a realistic situation. I have in mind typical problems these days in Silicon Valley where I work. Maybe you're starting up a company or several companies are starting up in using AI to automate and optimize supply-chain management. Each company is trying to figure out how to strategize, how much to invest, what the technology should be. In a case like that, it's not at equilibrium. The companies don't know what they're going to face. They don't know how the technology will work out. They don't know who the other players are going to be in this business. They don't know what the reception is going to be for this new technology. They don't know what the regulations are going to be, the legalities, the shipping arrangements. And yet they're in a position where they have to ante up maybe hundreds of millions of dollars just to sit at this casino table and play. Now, an economist would say that's not a question of probabilities. It's a situation where the firms face what economists call *fundamental uncertainty.* As Keynes put this in 1937,[4] "the prospect of a European war . . . the price of copper in three years' time . . . About these matters there is no scientific basis on which to form any calculable probability whatever. We simply *do not know*." Of course our firms know plenty, but if they want to define a real optimization problem or calculate probabilities, they simply do not know.

FUNDAMENTAL UNCERTAINTY

Realistically, agents not identical, don't know other agents' circumstances or likely behavior; or situation complicated and not fully known.

There is *fundamental uncertainty:*

- Therefore "the problem" is not well-defined;
- Therefore "rationality" is not well-defined;
- Therefore "optimal" behavior is not well-defined;
- Behavior and outcome unlikely to be in equilibrium.

Slide 2

There's a syllogism here. If a situation is subject to some high degree of fundamental uncertainty, the problem it poses is not well defined—you can't express it in clear logical terms. If the problem is not well defined, rationality is not well defined—there can't be a logical solution to a problem that is not logically defined. There is then no optimal solution, no optimal set of strategies, for agents and players—optimality is not well defined. So if you simply don't know what you're facing, rationality isn't well defined and optimal behavior isn't well defined. So where an economic situation contains significant fundamental uncertainty, if we are rigorous it can't be credibly reduced to a rational, deductive economic model. And where rational, deductive behavior is not well defined, any outcome is likely to be temporary and not in equilibrium. I remember sitting in Santa Fe in the early days and saying to Ken Arrow, "We should do something out of equilibrium, Ken. We should do something that has fundamental uncertainty. Don't you

4 Keynes, J.M. 1937. The General Theory of Employment. *Quarterly Jnl Econ*, 51, 209–33.

think that's important?" And Arrow, who was our godfather and arguably the top theorist in the world, looked at me and he said, "I realize these are huge problems," he said, "but what can you do about them? What can you do?"

In the face of real, fundamental uncertainty, the standard approach gets stalled.

This is where we were in the fall of 1988. And we talked a lot about it.

But something had resonated from the previous year, and it came as an inspiration from a modest, diminutive man. In 1987, I had a house the Santa Fe Institute had rented for me with a housemate who turned out to be John Holland. I had no idea of what he did; I had no idea that he was a large figure in computer science. I'd had a lot of discussions with him late at night over beer, and we talked a lot about John's passion for teaching computers to play chess or checkers. A month later, when we held this big symposium, John gave a talk the first afternoon. I listened and he was on about genetic algorithms and classifier systems and how to teach machines to get smart playing board games—how to write algorithms that could change their code and get smarter as they develop. We now call this *evolutionary programming*. John had all this down and was describing it and I'm sitting there thinking, there's something here we need to pay attention to. And the nearest I got to figuring out what it might be was just a hunch. What John was saying seemed to me immensely important. I sat there thinking, if John Holland is the solution, what is the problem?

[LAUGHTER]

John—and it took me quite some months to realize this—was pointing out two things: that people—or, if you like, automata, algorithms—*can* and do act in situations that are not well defined. This was mind blowing for us in the social sciences. Of course we do that. We enter contracts like marriage—not well defined. We decide we're going to get into Buddhism, but we don't know what that is. We do these things all the time, and John was pointing this out, but I'd been overeducated. Second, in not-well-defined situations, John said, in various ways people try to make sense: they experiment, they explore, they adjust, they readjust, but not just in terms of having some wondrous mathematical model of the situation and updating a parameter. They form a hypothesis, maybe they have multiple hypotheses or ideas about the situation they are in, and they put more belief in the ones that work over time and throw out hypotheses that don't.

This resonated with me, at least a bit later as our program progressed. I was working for a large bank in Hong Kong, in their Analytical Division. I had a team and we were trying to crack the foreign-exchange market at three in the morning when all the play was happening. We were using what now would be called machine learning. I noticed that the actual traders had hypotheses: "Oh, such and such has just happened, the central bank of New Zealand has just come into the market. But hang on, I think the Chinese are going to do that. And if the Chinese central bank does this, then that's going to happen." But then something else would happen. They'd throw out the hypothesis they were using and take in others. John had been explaining all this to us and the breakthrough was that I realized we could do this in economics. We could look at problems that were not well defined, whether there's fundamental uncertainty, problems where there is no equilibrium, and we could unleash agents that could explore, take up hypotheses, throw them out, and generate new ones and get smart.

We would have to use computers of course to track all this, and they would have to be based on the behavior of multiple "agents"—artificially intelligent computer programs that could have multiple hypotheses and use or change these. There was no name for this in 1990. And people were already writing programs that had multiple elements following rules and getting smarter. Not long after that the approach came to be called *agent-based modeling*. Agent-based modeling *emerges* from this approach as a natural technique.

Let me come back to our idea of an economic approach. We could model agents who had disparate ideas or hypotheses about the situation they were in. They could base their actions on these, and learn which hypotheses worked and which didn't, getting smarter over time.

Notice one thing here: these very actions of agents' exploring, changing, adapting, and experimenting further change the outcome, and they'd have to then re-adapt and re-adjust. So, they are always re-adapting and re-adjusting to the situation they create. This behavior is at the heart of complexity. I remember being asked in the middle of a talk like this, "Professor Arthur, how would you define 'complexity'?" I could have talked about elements reacting to the situations those elements create. But I had a better definition, "Have you ever had teenage children?"

[LAUGHTER]

Here's what I'm saying. In general in a difficult or novel economic situation, there's no optimal solution. Agents are coping, exploring, adjusting, experimenting, whatever. But that very behavior changes the outcome and then they have to change again. It's a bit like surfing, where you don't know where the wave will go next and you're adjusting and readjusting to stay in the green water.

Complexity economics is basically seeing the economy not as a machine, balanced and perfect and humming along, but as an ecology.

This lands us in a world where forecasts and strategies and actions, whichever you're using for that sort of situation, are getting tested within a situation—I would like to call it an *ecology*—which those forecasts and strategies and actions create. And, if you look at it that way, that's the essence of agent-based models. Complexity economics is basically seeing the economy not as a machine, balanced and perfect and humming along, but as an ecology. New strategies, new things are coming and going and striving to survive and do well in a situation they mutually create. We can describe this algorithmically, but not easily by equations, not just because the situation is complicated to track but because new behaviors and categories of behavior are not easily captured by equations.

This approach began to clarify itself in Santa Fe in the early 1990s, and there were other pods of researchers doing parallel work elsewhere. By the end of the decade the approach needed a name. In 1999, I was publishing a paper in *Science* on complexity and the economy. The editor called me from London and said, "You need a name for this approach." I said, "No, I don't." "Yes, you do." "No, I don't." He prevailed, and standing in a corridor on a landline I called it *complexity economics*. I thought it should have been called *nonequilibrium economics* afterwards, but it was too late. It's been locked in.

By the way, I'm not saying here that human agents in a particular situation are algorithmic, that they're walking around like robots. It's more subtle than that. You can model people learning in new situations where they don't know. The outcomes likely will not be in equilibrium. So people adjust again. The outcome changes again. More readjustment. The situation may eventually settle or it may show perpetually novel behavior.

And one other thing: with algorithms or computation, you can model agents *acting* as well. This is because if you describe something algorithmically, you can allow verbs. Equations allow only quantitative amounts of things—nouns. With algorithms you can directly use verbs: agents can buy, sell, change their minds, throw things out, create new things. These are actions. Verbs. So this new economics doesn't have to be about the levels of prices, or interest rates, or the amount of production, or levels of consumer trust. In complexity economics you get real verbs—real red-blooded actions—happening, and the state of affairs they produce could be complicated and real.

WHAT DIFFERENCE DOES COMPLEXITY ECONOMICS MAKE?

I've given a picture of how complexity economics came to be. Now I want to go back to our original question and ask what difference does it make? Complexity economics I believe makes practical differences to economics, and also a difference to how we understand the economy. I'll point quickly to four specific areas.

1. The persistence of financial crises. Standard economics has become good at dealing with recessions and preventing full-scale depressions, but if you look at financial crises in developed countries, these haven't abated in the last fifty years. If anything they have got worse. And when they hit, they cause serious hardship and disruption. Contrast this with the number of fatalities to passengers on jet aircraft. Around 1950, this was about forty-five fatalities per million passenger miles. Now it's down to four or five, close to zero. Passenger flying, seismic codes, cardiac procedure deaths have all improved greatly in the last fifty years. Your overall health has improved; your pet's health has improved.

What hasn't improved is the persistence of financial crises. In Russia's big bang around the early 1990s, Russia decided to go capitalist, but the thinking was equilibrium based. Old socialist planning equilibrium. New capitalist equilibrium. What might happen in between was fuzzy, not understood. In reality a small group of private players took control of the state's newly freed assets for their own benefit and industrial production plummeted. You can see the results of this in the vicious mortality statistics Russia suffered for five or ten years after, because they got it wrong. Similarly in 2002 California freed up its electricity market, and a small number of outside suppliers manipulated the market to their own profit. The state's finances were put in jeopardy. And we all remember the US subprime mortgage collapse of 2008, where exotic derivative products and negligent oversight caused an unstable structure to collapse spectacularly. Each of these systems was manipulated or "gamed," and all broke down. Nearly always when you see a financial crash, you'll find a small group exploiting part of the economy; this causes wider fissures and failures and may eventually lead to a collapse.

After the financial crisis of 2008, in England the Queen famously asked, "Why didn't economists didn't anticipate this disaster?" It's a good question. Some did, of course. A few said, "We have this weird mortgage-backed securities market in New York and I don't feel good about it." But overall the profession does not excel at seeing trouble ahead and unlike engineering disciplines it doesn't have a branch of forward-looking failure-mode analysis or of post-crash forensic analysis. Why is this? The reason, I think, is subtle. If you *assume* equilibrium, which economists do, then by *definition* cascades of collapse cannot happen. It's like *assuming* an engineering structure is in stasis, therefore it can't collapse. Similarly at equilibrium there's no incentive for any agents to diverge from their present behavior—that's the definition of an equilibrium. Therefore, exploitive behavior can't happen. A subtle, muted, unconscious bias precludes ideas of exploitation or collapse.

Complexity economics, by contrast, sees the economy as a web of incentives for novel behavior. It sees the economy as open to new behavior, and that might turn out to be exploitive behavior. And it starts to ask, we are in this situation now, or we have this policy system in place now, what's the likely response going to be? Can someone exploit the system? Can some group game it? These are healthy questions to ask. The economy is not a closed static equilibrium system; it is a system perpetually open to novel behavior, and complexity economics forces us to keep this in mind.

Nearly always when you see a financial crash, you'll find a small group exploiting part of the economy; this causes wider fissures and failures and may eventually lead to a collapse.

2. *Awareness of the propagation of change.* The second thing complexity economics can help a great deal with is an awareness of the propagation of change. If you're looking only at equilibrium, there isn't any change, and so there isn't any propagation of change. But, as we heard in a previous talk from Matthew Jackson, banks—or people, or economic agents in general—can and do affect each other, and they do this through networks of connection: trading networks, information networks, lending and borrowing networks, networks of disease transmission. In these networks, events can trigger events and failures can cascade and cause disaster. Banks in distress can infect other banks. If I'm a bank and I'm in trouble and fail, then I pass on distress to my creditor banks, who may pass on distress to their creditor banks. Stress can cascade through a system like this.

Equilibrium economics traditionally didn't cope with this sort of thing very well. It traditionally assumed that firms were independent, and so changes would be independent, and so their sizes and aggregate effects would be distributed normally. One of the great things about complexity is we now understand this sort of thing—interconnections between firms and agents and how these work—much, much better. We understand that individual banks and firms are not independent. Events with one bank can trigger events in other banks in the network, and so systemic risk—overall risk to the system as a whole—is not the summation of independent events, but it is reflected in domino-style avalanches of various sizes and duration. This gives higher probability to large disturbances than normal distributions would predict. And it yields power laws of various kinds. Propagation lengths, sizes of firms, and magnitudes of cumulated events are distributed not normally but according to some law where lengths or sizes or magnitudes fall off logarithmically, much like earthquakes do on the Richter scale. This sort of thinking is now crucial in modern finance and derivatives trading, and understanding gained from complexity in general and network behavior in particular excels in this area.

3. A strong link with political economy. Complexity economics also connects with the insights of earlier political economists and this gives both a depth and a validation to its insights. And it opens up a venerable literature to modern perusal. This isn't obvious, so let me explain.

I realized quite a while ago that in economics there are two large groups of problems. See slide 3.

You could call one group *allocation problems*. That's international trade theory, strategic outcome theory or game theory, general equilibrium theory, etc. Those are all well understood. And they have been

TWO GREAT PROBLEMS IN ECONOMICS

- **Allocation in the Economy**
  Quantities: General equilibrium, international trade, game-theory outcomes . . .
- **Formation in the Economy**
  Processes: Of econ development, discovering novel technologies, structural change, arrival of new institutions, temporary phenomena like bubbles, crashes . . .

The former is mathematizable; the latter, not.

Slide 3

mathematized, heavily mathematized. They are about quantities and preferences and prices in balance, and this gives itself to the equating of things—to balance and to equilibrium.

The other group of questions are ones of *formation*. How do economies come into being? How do they develop? How does innovation work? Where do institutions come from? How do institutions change things? What really is structural change, and how does it happen? These latter questions of formation are all ones that complexity economics can look at. Formation precludes equilibrium, and it's about how structures or patterns form from simpler or earlier elements. That's what complexity is about, par excellence, and complexity opens the door to rigorous study of all these questions of formation.

Formation is not new just because complexity has discovered it. It was studied in earlier economics, particularly within the older schools of classical economics, political economics, and Austrian economics. These differ somewhat, but together they provide a view of the economy as not necessarily being in equilibrium, always being in process, always subject to historical contingency, always creating and responding to the rich context it creates. None of this can be mathematized easily into balance equations, but it can be studied via the methods of pattern formation, network analysis, and algorithmic models heavily used in complexity economics. Not only can we can meet up with these older and wiser approaches to economics and learn a lot from them; we can also help make rigorous their insights, and thereby see the economy not as a static system, but as one in process, always exploring, and always forming itself from itself.

4. We can model reality more rigorously. Complexity economics sometimes uses mathematical equations, but more often its models are complicated, so to track things properly we have to resort to computation. This gives us if used properly an important advantage. Mathematical models in economics are forced to be kept simple, largely to allow pencil-and-paper analysis. They typically track rates of how aggregate variables influence aggregate variables—how average wages, say, are related to unemployment rates. And they need to be based on simple assumptions: identical agents that are rational and behaving in an equilibrium setting. With algorithmic models we can include as much detail as we want, to an arbitrary depth, and so we can free models from the tightness and inaccuracy of unrealistic assumptions. We can have agents who are realistically diverse; we can include details of how they interact, of the institutions that mediate these interactions, details of networks and interconnections. Of realistic behavior. We have to be careful here, as with all modeling it's easy to lapse into throwing in useless or inaccurate detail. But done properly such detail adds realism—and thereby rigor.

Not surprisingly, this new realism is giving us much better understanding of the processes of the economy, how, say, the 2008 crash happened, how diseases actually transmit, how economic development actually takes place. More precise detail gives sharper resolution to the instrument of economic analysis and we see things that would not be visible otherwise. I don't think there is an upper limit to what we can learn here.

CLOSING THOUGHTS

Let me close with some thoughts here.

Complexity economics brings a different view of the economy—a different understanding to the economy. In standard economics, problems are well-defined, solutions are perfectly rational, outcomes are pure, possibly artificial in a way, but above all they are elegant. And by assumption outcomes are in equilibrium. Loosely, the economy *is* in equilibrium. Standard economics in a word is orderly. To borrow architect Robert Venturi's phrase, it has "prim dreams of pure order."

In complexity economics, agents differ and in general lack full knowledge of each other and of the situation they are in. Fundamental uncertainty is therefore the norm; ill-defined problems are the norm; and rationality is not necessarily well defined. Agents explore and learn and adapt and open to novel behavior. Outcomes may not be in equilibrium. Complexity economics deals with historical contingency, path dependence, and to quite a degree indeterminacy. It is organic, with one thing building on top of another. It is a cascade of events triggering other events, and so it is algorithmic. To borrow another phrase of Venturi, it's full of "messy vitality." In philosophical terms, standard economics I would say is Apollonian, pure and ordered; complexity economics is Dionysian—it is structured and generative, with just a dose of wildness.

To go back to our original question, is this economics just a bolt-on to standard economics? Absolutely not. Nonequilibrium includes equilibrium, so it's a more general theory. It's relaxing the restrictions of equilibrium economics—well-definedness, rationality, identical agents—and thus generalizing standard economics.

Does this go with the *Zeitgeist*? I'd say yes. The sciences from roughly 1900 to about now have gone through a shift, I believe, from order, determinacy, deductive logic, and formalism to formation, indeterminacy, inductive reasoning, and organicism. There are many reasons for this. Part of it is the rise of biology and molecular biology as serious sciences and computer science and new types of mathematics. And, by the way, one of the symptoms of this is complexity itself. Complexity is not the cause of this viewpoint; it is more the outcome.

Let me finish with a rather beautiful allegory that the economist David Colander has given. Colander says, imagine that around 1850 all economists are gathered together at the foot of two huge mountains and they decide they want to climb the higher mountain. But both mountains are in the clouds and they can't see which one's higher and which one is not. So they decide to climb the one that's more accessible, the one that they can get their equipment through to, and the one where the foothills are better known. This is the mountain of well-definedness and order. They get to the top of that, and once they're above the clouds, they declare, "Oh my God. This other mountain, the one of ill-definedness and organicism, is higher." That's the mountain we are starting seriously to climb, and much of the journey started at Santa Fe.

[APPLAUSE]

AUDIENCE MEMBER I work for a hedge fund. About financial crises: you said there's always a small group of actors that create the crisis. In the case of the 2008 financial crisis, who do you think that group constitutes?

W. B. ARTHUR I don't know this history very exactly, but a number of people in the audience do know it well and could answer this better than I could. What really caused this crisis, I think, were people who allowed good credit ratings to be given to what turned out to be worthless products and derivatives.

What I would point out is this: I'm trained as an engineer and after the crash I read a lot about failure mode analysis in aircraft. What you learn is that there's always some tiny crack that appears, some tiny event that triggers another event that might trigger another event. Nearly all the time, nothing disastrous happens. But sometimes these events triggering events lead on and eventually disaster happens.

Why should complexity economics be any better looking at things like that than static equilibrium economics? Complexity economics is a way of looking at the economy based on events triggering events. Static economics equilibrium, looks at, well, equilibrium. It doesn't give you any feel for this triggers that, that triggers this. An economics that's built on stasis isn't biased toward seeing cascades of collapse.

AUDIENCE MEMBER: I'm a value investor. There is an underlying assumption in modern portfolio theory that people are rational. That seems to be the same assumption in traditional economics. Are there similarities between modern portfolio theory and traditional economics?

An economics that's built on stasis isn't biased toward seeing cascades of collapse.

W. B. ARTHUR: I spent time with Blake LeBaron, who's in the room, and John Holland and Richard Palmer, looking at what's called asset pricing under conditions where people aren't necessarily rational. They're exploring and trying to find out how the market works as they invest. Sometimes standard economic theory works very well. Sometimes you can think in terms of rational players, think of probabilities of what earnings might be, and you could think of lowering risks by diversifying among stocks that are not that correlated. That's portfolio theory and it's fine. When our team in Santa Fe looked at asset pricing we used a computer model with "artificial" investors; these weren't "rational," and didn't know anything at the start (say, as in Alpha_Zero) but could learn from scratch what behavior was appropriate in what market situation. We found that realistic market phenomena emerged from our model: technical trading, bubbles and crashes, autocorrelated prices, random periods of high and low volatility. These were emergent phenomena and they occur in real markets.

I was talking about this Santa Fe artificial stock market one time in Singapore and a hand went up in the audience. "You've said in this talk that ninety-eight percent of pricing can be established by standard economic models and about two percent of pricing can be explained by these events triggering other events, bubbles, crashes, and so on. So we're only off by two percent. What do you think of that?" I said, "Well, that two percent is where the money gets made."

[LAUGHTER]

You could equally argue that from outer space the oceans and the world are perfectly at equilibrium spherically, within maybe fifty or a hundred feet. But the nonequilibrium part is where the ships are. And that's what counts. It's the startups that make it. It's people who are lucky, the ones who understand how the nonequilibrium patterns work. You can't model that with equilibrium.

AUDIENCE MEMBER Could one argue that complex systems like emergent phenomena are a form of equilibrium? Maybe they don't last for a long time, maybe they change slowly, but there is still some kind of equilibrium.

W. B. ARTHUR My answer to that is no. Not really. You can do that, I suppose, but it's a case of paradigm stretching. You can think, as I said, of cars with an equilibrium flow, of cars going down a freeway. And you could say that, in a situation like that, if the cars get quite dense, then an emergent phenomenon occurs, there might be a traffic jam. A car slows up for some reason. Maybe some animal runs on the road, cars behind it slow up, and quite soon there's a traffic jam, an emergent phenomenon. Now, and this is subtle, you could define a stochastic process that was in equilibrium, a stationary stochastic process of car flow, and you could set that stochastic process up so that every so often it fetches up occurrences like traffic jams. You've preserved "equilibrium." But I'll give you my reaction to that—Yeecchh!

[LAUGHTER]

Why not just relax and say, "Equilibrium's wonderful. I'm not against it. It's given us a huge amount of economics. It's given us wonderful insights, but there are situations that are not in equilibrium," rather than try and say it's all equilibrium?

We need to admit different types of phenomena besides just equilibrium ones. We need to allow that the world is messy, always unfolding, always giving us new things.

AUDIENCE MEMBER I'm from a national laboratory. How do you build a common knowledge with the new kinds of analytic frameworks that we have for nonequilibrium so that people easily can understand what they're seeing when they're looking at analyzing these kinds of verb type of measures and ways of framing the problem rather than the nouns? We haven't figured out how to do that easily for just kind of the general common knowledge other than specialists.

W. B. ARTHUR Earlier today Eric Beinhocker gave a brilliant talk. He said we need to look at things—in the economy, maybe in politics, or situations in general around the world—using different vocabulary, different types of instruments. We need to admit different types of phenomena besides just equilibrium ones. We need to allow that the world is messy, always unfolding, always giving us new things.

I think the entire *Zeitgeist* is changing, and we're in the last stages of Enlightenment thinking. In the 1730s, people thought everything was mechanism and everything was right and everything was properly ordered, and "All nature is but Art, unknown to thee; all Chance, Direction, which thou canst not see; [...] all Partial Evil, universal

Good." That's from Alexander Pope's 1733 *Essay on Man*, basically saying everything is ordered, everything is understandable if only we could look into it and understand its mechanisms or "art." And that's changed. I don't know whether it's world wars, or quantum physics, or biology. We're looking now much more at the world as swirling and changing and contingent and not fully knowable and organic, with one thing building on top of another. Santa Fe Institute would be in a very good position to say, if there is a new way of looking at the world, what exactly is it? Where is it coming from? What evidence do we have? Why do we think this way now, when we didn't do this forty or fifty years ago?

Our outlook has changed and I believe we will be talking a slightly different language. If you think everything should be perfect, and ordered, life gets brittle. "Oh, my life isn't perfect. Things are out of order. I haven't done this properly. I haven't done that." We get nervous. If we say, "It's pretty good but could be better," then it's one damn thing after another, which is a definition of life, and we're okay. We will adapt. Then people relax, the polity relaxes, social ethics relax.

I think we're missing a way of looking at the world where we accept life. I was influenced by Robert Venturi's 1965 book *Complexity in Architecture*. He meant complication. But he made a huge distinction between what he called the "prim dreams of pure order" of the Bauhaus, all these geometric perfect things, versus the ambiguous, the contingent, and then a whole list of things. "What I'm for is messy vitality," he said. I think if there's a message in complexity and in its *Zeitgeist*, it is, let's cool it on the order. Let's just relax into a world of some degree of messy vitality. For my money, complexity shows us what that world looks like. It's not optimal, but it's pretty damn good. And it's alive.

DAY 2: PANELS

Quirk or jaguar

SFI economics

ER, GOD, IS IT SHE?

— SCOTT E PAGE

In mathematics, **ergodicity** is a property of a discrete or continuous dynamical system which expresses a form of irreducibility of the system, from a measure-theoretic viewpoint.

6

COMPUTATION & COMPLEX ECONOMIES

[Panel Discussion]

Moderated by David C. Krakauer,
featuring Robert Axtell, Joshua Epstein, Jessica Flack,
Blake LeBaron, John Miller & Melanie Mitchell

DAVID C. KRAKAUER Yesterday we had an overview of different perspectives on what economics might be missing—and of course that could have been much longer. And today we will ask that question through three different perspectives that are quite dominant at the Santa Fe Institute. One of them is the physics perspective. One of them is the computational perspective. And the other one is the sort of biological, organicist perspective.

Physics basically says we'll understand the universe because we're going to extremize a function. That's what physics does—optimization, some least-action principles. So, we're going to minimize the time, we'll minimize the energy, we'll minimize the distance. We're looking for those functions that we're trying to extremize and we're going to represent them in some minimal form and we're going to live with approximation. That's physics.

And that's quite different from this panel that says, no, there might not be a function that you're extremizing. That's not what we're doing. Actually, you should think of the system in terms of rules with mappings from inputs to outputs. The computational framework introduces a very interesting notion, the notion of correctness—which is *not* present in the physics language. There is nothing right in physics, but there is something right in computation. And its foundations obviously don't come from the natural world, but from the study of

---

OPPOSITE *Economic Curiosity: Civet Coffee*

logic and mathematics. It's a very weird framework that doesn't come out of empiricism, but of a certain kind of armchair philosophizing that turned out to be very useful. So the minimality expectations can be dropped. You could have a very, very cumbersome computational system that produces the correct output. So that's unlike physics.

JESSICA FLACK I disagree.

D. C. KRAKAUER You disagree. Excellent.

And then the final framework would be the sort of evolutionary organicist one that Brian was talking about, which is thinking about these agents as a collective in an ecological setting, quite different from physics and from computation. So that's the big framing.

Each panelist will make a five-minute remark, hopefully provocative. We don't want this to be a dull panel. We want it to be exciting. And if you have a strong opinion on what they say, just interject. We're quite happy for you to shout out and disagree with them. And so, with that, I'll allow them to self-police and I'll ask the occasional question if you all sort of diffuse. Okay. John, you want to start?

JOHN MILLER Sure. Thank you. I promised David I would give a four-minute-and-fifty-seven-second introduction to all of economic theory. So, this is going to be done by metaphor. It was inspired by this great book called *The Ashley Book of Knots*, which was written in 1944 by an artist and sailor named Clifford Ashley. He catalogs over 3,800 different types of knots. But when I was reading this book—not all 3,800—I came upon his definition of the knot, which I thought was just great, namely, "any complication in a length of line," which is useful, concise, and poetic. This got me thinking about economics. The metaphor here is that we have all these ropes and knots, some very practical, some ornamental, some often tangled messes. And you know, economic theory has gone through different phases.

In the early phase of economic theory, basically, they just cut out the knots and paid attention to each knot individually. So you have Adam Smith and the pin factory. Then, over time, and obviously most notably in the last eighty years with the rise of neoclassical economics, we just assumed away the knots and paid attention to the straight lines. More recently, we've actually gone back to paying attention to the knots.

One example is the rise of behavioral economics over the last thirty years. These guys have accumulated this huge collection of knots. How huge, you might ask. Well, I went to Wikipedia this morning to figure out how many heuristics and biases they've cataloged. I stopped counting at 190 and that took me from the ambiguity effect to the Zeigarnik effect. So the current status of behavioral economics is not unlike particle physics before Murray Gell-Mann and George Zweig came around, where we had this ever-increasing zoo-full of particles or effects.

Obviously the physics issue was solved by the notion of quarks or aces, where they took this huge zoo and suddenly simplified it in a really useful and productive way. In economics, and certainly in behavioral economics, I'm not so optimistic that that's going to happen. I mean, it'd be great if somebody came along and came up with the quark equivalent for behavioral economics. If they do, I hope they call it the quirk because it's a kind of behavior . . .

[LAUGHTER]

. . . but I think economics has gone between these two extremes, one where we vastly simplify the world and the other where we have a very intense and isolated investigation of the knots. Computation is this middle path where we can really start to explore both of these elements, where we can look at economics and behavioral phenomena both with the lines and the knots and really try and start to generalize.

It's like old-home week up here on the panel. So everyone has been around—most of us, anyway, so that'll give anyone immunity if they want it—doing this for about thirty years. The big secret, and I know we have a lot of financiers and investment people, so try and keep this quiet, but computation has gotten a lot faster in the last thirty years.

[LAUGHTER]

In the early years we were able to do a model, get some insights, and you might think that the increase in computation might cause the behavioral economics problem that now we're going to get too many insights. But I think the answer is actually the opposite: that, with enough computational power, we can run enough examples and even use clever algorithms on top to start to unify this much more interesting area. Let me stop there.

BLAKE LEBARON I thought I would actually just give a quick one-minute summary of what I'm going to speak on, because I've got all these distinguished agent-based people up here. I will sort of cover agent-based finance in five minutes as best I can with a quick summary of what's been done in the past and then where it's going. Hopefully four minutes on that.

Obviously, as many of you heard, one of the earliest agent-based models was the Santa Fe stock market. My co-authors—a great team of Brian Arthur, John Holland, Richard Palmer, Paul Taylor—we all worked on this. Started it here, actually, in the late '80s, early '90s, and it kind of burst into a big industry of other people doing the same thing.

Heterogeneity is a bit of a nebulous feature in complex systems; it is a bit of a nebulous feature in finance.

I would say, interestingly enough, anthropologically about 80 percent of the research in agent-based finance goes on in Europe, not in the US. We replicated a large number of features. That was our kind of big deal: things like fat-tail return distributions, persistent volatility, volume volatility, dynamics, PE ratio movements. All of this stuff we were very good at. Over the years, we've gotten better and better and better at it. All of that was of course completely endogenous in these markets, which kind of drove the traditional finance economics world a little nutso. As we've talked about before—and Brian mentioned in his talk—emergent features do drive people a little nuts, because they don't quite know what's going on.

Two interesting features that I've done on a lot of later markets: one, I've been looking a lot at forecast surveys, what people think the market might be doing. If you're not doing rational expectations, you should at least be looking at what people's expectations are. What we know from that is they go the wrong way. Generally, people predict markets are going up when prices are high and they predict them going down when prices are low. Agent-based models do that quite handily. But there's also a lot of very interesting heterogeneity in the data, too, which I'll get to in a sec. Another big stylized fact, which is just part of the design of these markets: generally, flows—into various managed funds or into any kind of fund—chase previous returns. Investment flows chase previous performance. This is just something that's the design of the market's agent. Most good agent-based markets have that in their design. It's not even a question, not part of a lot of traditional models. So, that's the landscape of this. I think that we obviously did pretty well at all of these things.

For the future, I think a lot of different pieces are important. I think heterogeneity is key in these models. I've been saying this for a long time. In finance, heterogeneity maps directly into liquidity. Heterogeneity is a bit of a nebulous feature in complex systems; it is a bit of a nebulous feature in finance. Many of you, of course, know what it is. It deals with the ease of making trades. Heterogeneity is part of that, and so understanding it in a deep sense is very important. I think we have not been very good at modeling multi-asset markets and you need to understand that to understand liquidity. So we're pushing hard. I'm pushing hard. That's sort of like the ultimate shot there, to understand things like crowded trading and these sorts of phenomena in a multi-asset world. That's something we've not done a very good job at.

But remember, emergent features are not just variances, but also covariances and correlations. I think many correlations in financial markets are purely emergent features, not part of the underlying framework. Brian mentioned this the other day; he used the term "ecology of traders." That's totally relevant to what we were thinking about. Along the way it's now become super important. We were always sort of thinking about that—the presence of differing types of funds, passive versus active, all of these semi-passive ETFs, all of that landscape has come up in the last twenty years. Now, you know, maybe we should better turn our tools to understanding that landscape.

Last thing: "it's the data, stupid." Data, of course, changes a lot of things here. John was right, of course, on compute power. I don't disagree with you, either. Finance is a little tricky. Some of the data's hard to get; some of it's proprietary. I've been in many fights and discussions, trying to get access to appropriate data to do micro-level validation. One area that's probably going to be important because we're seeing it more often is these generative systems, where you kind of hand generation of data to some deep neural network, hoping that you build a system that basically builds the features of the data so we can all use it. It's not crazy. The US census is thinking about doing this, too, or has played with those ideas. Also, finally, using just the big sets of machine-learning tools in many different ways, determining trader type, and actually coming back to the old question of what markets look like when they're populated by traders who really are trying to learn against each other, closer to AlphaGo in spirit than maybe other standard agent-based models. That was actually part of our original goals.

J. FLACK Perhaps one clarifying remark before I make my comments, just so you can sort of situate my position on this panel. David started by saying that there are these three distinctions at SFI: the physics, the computational, and the organic, and I think that's basically right. I'd add a fourth, which is the idea that the system itself might be computing. The computational perspective, which is definitely represented on this panel, is one in which you build generative models, and agent-based models, and so forth. And it's not actually necessarily at odds with the idea that the system is computing. I think that will come up in discussion. But I want to stress that what I am going to be talking about is the system doing a computation, and that's very, very important to understand.

Let me start by saying, then, that the issue I want to put on the table for discussion is information. A lot of you are probably thinking, well, that's not a new concept. Many of you might respond by saying that information is incorporated into the models that we build. We take into account how agents process information, and that's all true. But the fact remains: we do not have a theory for how you get micro-to-macro maps and information-processing systems. We have no theory for that. So, to that end, we've been working on building a theory of what we call *collective computation* to account for how macroscopic states with functional consequences for individuals or agents arise from microscopic interactions in information-processing systems.

The point really is to explicitly allow for subjectivity, which comes from noisy data and noisy or erroneous information processing by components, to understand how this factors into what gets produced by the system. Okay. So, the theory of collective computation aims to predict. A lot of times we're wondering, well, what can these theories predict? We're trying to predict something like the precise value of the output. That's part of it, but what we are *really* aiming to predict are mechanistic solutions to certain kinds of challenges. And some examples of that might be a dynamical property, like the distance from a critical point; a feature of social structure, like the distribution of power inequality; maybe something quite simple like, in a neuroscience experiment, a monkey has to make a decision about which direction dots are moving on a screen.

. . . we do not have a theory for how you get micro-to-macro maps and information-processing systems.

So, a very simple decision at the whole organism level. Another thing about collective computation, of course, is that we're aiming to address how the computation is refined; over evolutionary learning time gets better—that notion of correctness David mentioned in his introduction.

One of the main tenets of collective computation is that the subjectivity I mentioned at the beginning can be overcome by aggregating component estimates. That's kind of a collective intelligence–like perspective. Then, depending on constraints in the system and adaptability issues, good estimates can be achieved by balancing investment in something we call *information accumulation*, which is how well the agents actually estimate regularities in the environment against the algorithms that we use to aggregate information. So, if your agents are not particularly good sensors, there are—in the behavioral economics case—a lot of biases. Maybe you build aggregation mechanisms that compensate for some of those biases.

A second tenet is that the constraints imposed by considering how information is processed from the system's perspective are super-important. I mentioned that at the beginning. We have to take into account that that plays a big role when it gets produced. So if these tenets are correct, then this implies—and this comes back again to some issues that may be raised in this panel—that we can't deduce the strategies individuals are using. We have to get them inductively from data because they're making estimates about how the world works. We have to understand how they're doing that in order to understand what they've seen and how they choose the way they behave.

So, four concepts that collective computation brings to the floor are: first, ground truth, this notion that the Eiffel Tower has a height that we can measure. An effective ground truth is the idea that we might all collectively agree what the height of the Eiffel Tower is, but it might not be right. What's important is that we agree what the height is.

A second important concept is the idea that information can be collectively encoded in a circuit or a network and not reside in the heads of any individual actors, necessarily, or be reducible to that. So, for example, in the collective intelligence literature, Galton asks, what's the weight of a steer? He took the average of the average estimates of individuals in the crowd. That's a very simple aggregation mechanism and it's not really collectively encoded.

Third, a lot of the stuff we work on is where our aggregation mechanisms are, where the information is in the network—it's in the interactions.

And, finally, this idea that outputs, the things we're interested in, are collectively computed or constructed by the system components. You can think of the output as essentially the result of collective dynamics, for example, like electing a president. Of course the president has some intrinsic abilities that we think might make him or her a good president, but the fact is how we respond to that person influences their performance. And that can have a slow timescale. We need to take those into account.

I have some comments about efficient and inefficient markets, but I will make those in the discussion.

ROBERT AXTELL Okay. David encouraged me yesterday in particular to not be particularly agreeable, so I'm going to start off by disagreeing with him. He quoted Adam Smith and Darwin and Keynes and then eventually got to Newton. And he basically said, in so many words, that economics needs to figure out the motions of animated things. Using

AGENT-BASED MODELS FOR ECONOMICS

Robert Axtell

Computation is revolutionizing all of the sciences, but it seems to me that the economics profession has been relatively laggard in its adoption of modern computing ideas. On the one hand, numerical economics is the conventional approach to solving so-called computable general equilibrium (CGE) and dynamic stochastic general equilibrium (DSGE) models for policy purposes, but such efforts are subject to the same criticisms as the neoclassical theory that underlies them: they use rational agents while people behave in a variety of ways that are demonstrably not rational; agents are often assumed to be homogeneous; networks between agents (people, firms) are neglected; and disequilibrium dynamics/paths are often not available since the models are solved for equilibrium.

But as a practical matter, these approaches use very little of the hardware now available on state-of-the-art workstations. For instance, such models often employ quarterly data—perhaps disaggregated by some tens of sectors—so they have no large-scale storage or memory requirements. Nor are the tremendous advances in display technologies, including GPUs, typically utilized by economists pursuing numerical strategies. While large numbers of difference or differential equations may be solved, or complex mathematical programs run to accomplish optimization, these primarily exercise the CPU and its cores and caches while leaving much of the rest of modern machines quiet. Other more theoretical approaches to computational economics, such as the perceived need by some to do such computations without real numbers—so-called "computable economics"—is also an intellectual cul-de-sac and offers no progressive research program for economists.

Rather, it is precisely the kinds of computing that were foundational to SFI thirty years ago—artificial life, cellular automata, and software agents—that hold the most promise for revolutionizing how economists employ computing, in the guise of agent-based modeling. By specifying agents and their interactions and letting macrostructure emerge, ABM is a kind of macroscope that can be used to study emergent social and economic phenomena, particularly the kinds of things that can have harmful impacts on people. For instance, the housing-market bubble that burst in the first decade of the twenty-first century was a macro-level phenomena that we have shown how to model using ABM.

. . . it is precisely the kinds of computing that were foundational to SFI thirty years ago—artificial life, cellular automata, and software agents—that hold the most promise for revolutionizing how economists employ computing, in the guise of agent-based modeling.

Similarly, ABM is the main technology for studying the health and economic impacts of

SARS-CoV-2 and the corresponding COVID-19 illness.

This is so because ABM facilitates the study of systems for which the mechanisms are known (to some extent) but there are not good data about comparable events from the past—thus the application of ABM to disease-transmission models, as well as economic-impact models for the novel coronavirus, a pandemic at a scale unseen in recent decades and therefore not something about which we have good data. ABMs make full use of modern computing by filling up all available RAM with software objects/agents representing individual economic actors, leveraging display and GPU technology when possible to both speed up calculations and display high-dimensional results, while utilizing all available cores to parallelize model execution.

It seems to me that ABM has the chance to move quantitative economics beyond its current mathematical phase, in which empirical relevance is defined in terms of econometrics/applied statistics, and more toward a bottom-up, decentralized, distributed view of the world in which big data plays an increasing role. ABM is a natural "language" for expressing economic-process models and a suitable basis for computation within economics, both practical and theoretical.

Further Reading

Nontechnical: "Economics as Distributed Computation," by Robert L. Axtell, 2003, in H. Deguchi, K. Takadama, and T. Terano, eds., *Meeting the Challenge of Social Problems via Agent-Based Simulation*, Springer: Tokyo.

Technical: "The Complexity of Exchange," by Robert L. Axtell, 2005, *Economic Journal*, 115 (504): 193–210.

the Newtonian paradigm is just a starting point, but we need to move beyond that. But I want you guys to think about this quote from a panel at the American Economic Association some time ago.

Here's what the speaker said:

> *For centuries after Newton, systems of differential or difference equations provided the model of ideal scientific explanation of dynamic systems. But the modern digital computer, with its very general capacities for representing symbol-manipulating systems, opens to us the same possibility for the time stream of a decision process by means of a computer program. Thus, the advent of the high-speed digital computer . . .*

—oh, that dates the quote a little bit—

> *. . . economics has acquired a theory-building and theory-testing tool that will enable it to handle far more detail of firms' behavior than could be treated in the past.*

He ends by saying:

> *With optimism we may even hope that the demands of the institutionalists for faithfulness to facts no longer seem irreconcilable with the demands of theorists for facts that are manageable. We will feel less constrained to believe in a particular kind of world just because it happens to be a world that is easily theorized about.*

Guesses who that is?

AUDIENCE SUGGESTS HERB SIMON

Of course! What year? '62. That's amazing. Sixty years is a long time ago, right? I mean, he's saying more than what David said. He's saying that we actually have today a technology which we can use to do what we want to do, which is basically to model these decision-making streams and to get answers that way. Now, at the time, his junior colleagues at Carnegie—Dick Cyert, Jim March—they would go on to publish what would be their book *A Behavioral Theory of the Firm*. It's a milestone book that recently had its fiftieth anniversary. But that book in particular describes a computational model of how a firm behaves, situated from the point of view of how decisions are made within the firm.

Soon after that book came out, Marvin Minsky invited Simon to go to MIT to give lectures on artificial intelligence. At the time it was called AI. That was Minsky's term, not Simon's. Simon had a different term: complex information processing. Anyway, out of those lectures came the great book *The Sciences of the Artificial*, which many of you know, and if some of you don't know that book, just read the chapter called "The Architecture of Complexities." It's extremely worthwhile to read. Well, it turns out that, what you might not know is that there are three editions of that book. And Simon wrote, in the third and final edition in '96, a special chapter just for SFI, and in it he goes through what he thinks are SFI's contributions to the ideas of emergence and etc., artificial life, etc.

. . . systems need a correct balance between exploitation and exploration.

He's broadly supportive of SFI's goals and motivations and work in general. By the way, as a footnote, let me just mention in the short time I have that Herb Simon never visited SFI. There's a great story about that. Doyne Farmer asked Murray Gell-Mann one time why Herb Simon never visited SFI, and it was a very uncharacteristic Murray answer to that, which I'm not going to repeat here, I think. I'm going to wait. If we have proceedings, I'll get Doyne to put that into proceedings. Some of you may know it.[1]

So, with this as background, I want to disagree with something that Brian said yesterday. Brian said basically that complexity economics was founded at SFI in the late '80s, early '90s. I think complexity economics was founded in Pittsburgh in the 1950s and '60s, and we have had a steady stream of it since then. The contribution of SFI has been large and, in fact, I would even suggest that the golden age of SFI and computational economics—complexity economics—was under the tutelage

1 Doyne says, "I once asked Murray why Herb Simon had never visited SFI. Murray got a funny look on his face, and said that he had once been on a panel with Herb, and that Herb was so smart that Murray found him really intimidating. My question was left hanging in the air. If you knew Murray, you will have some idea of how exceptional Herb Simon must have been!"

of Blake as the economics director in the early '90s. Since then, the SFI community has never really had a computationally oriented economics program. And that's a bit of a lacuna, given SFI's early start in the area and its niche. I'll stop there.

MELANIE MITCHELL Okay. I was a little bit worried about whether I had anything to say about computation and complex economies since I'm a computer scientist. I would have preferred computation *or* complex economies. But yesterday, listening to Brian Arthur's talk, I was inspired by his citation of John Holland as a relevant figure in this area. John Holland was a computer scientist. He died a couple of years ago and he was my co-advisor at the University of Michigan. Brian Arthur noted that John Holland focused on the need for agents to adapt to uncertain, ill-defined, changing situations, which is really the basis of a lot of what we think about at SFI. It's not only true of people, but it's true of companies, governments, insect colonies, brains, individual cells, and so on. It's really the foundation of adaptation.

When I was in my first year of graduate school at Michigan, I took a class from Holland called "Adaptation in Natural and Artificial Systems," which was the title of his book. A very odd title for a computer-science course. John Holland actually was the very first PhD in computer science in the US, which is maybe not too well known.[2] He viewed computer science as much broader than just computers. You know, these hardware objects that either are these huge refrigerator-size mainframe computers or these small things that now fit in our phones. But that it was in some sense writ large across all of nature. And this was the foundation of the course. The aspect of the course that stuck out to me in subsequent years was maybe the simplest but most profound principle of adaptation: that systems need a correct balance between exploitation and exploration.

This has become a buzzword in complex systems—the exploitation/exploration balance—but it's really fundamental to almost everything that we do. The idea is that you can exploit things that you already know,

---

2 Holland, whose dissertation was titled "Cycles in Logical Nets," received a PhD from the University of Michigan's new Communication Sciences program in 1959. This program became the "Computer and Communication Sciences Department" in 1968. https://cse.engin.umich.edu/about/history/

that you've already explored, or you can choose to look at new things, try out new things that you haven't explored yet that might be better than the things you've already looked at. Holland in his course talked about a formalization of this problem as a two-armed bandit, a slot machine with two arms. You can choose which arm to pull at each time step, and each arm has a different probabilistic payoff. So you can either observe the machine and then exploit the arm that has the seemingly highest payoff, or you can give trials to the other arm, which is maybe less explored, but might statistically end up having a higher payoff than the samples that you've already seen.

And the question is, how do you find the optimal balance of trials to give to these two arms? You can generalize this. Holland made an analogy between evolution as a multi-armed bandit, for instance, exploring many different kinds of genotypes or species. Each species that emerges corresponds to a pull on an arm and its fitness is a sample of the arm's payoff. Holland noted that evolution was in some sense optimally exploiting and exploring at the right balance, and could formalize this with genetic algorithms. In fact, the genetic algorithm wasn't developed by Holland to be a new computer-science, machine-learning technique. It was really a formalization of what adaptation was in the real world and Holland was able to derive mathematical properties of the genetic algorithm's exploitation-versus-exploration balance.

Now, it seems to me that a lot of problems of complex systems result from having a maladaptive balance. I was also on Wikipedia looking at lists of cognitive biases this morning and it seemed that a lot of our biases, like confirmation bias, availability heuristics, salience bias, our selective perception, etc., could be formulated as maladaptive exploitation-versus-exploration approaches. Or perhaps they're the best we can do given the kinds of constraints that we have. I'll pose this as a question to you: How much do we have to think about this issue—these various cognitive biases, these maladaptations, or these constraints—when thinking about economic systems? Do we have to worry about these kinds of psychological exploitation-versus-exploration issues?

Recent machine learning has really focused on this exploitation-versus-exploration issue, especially in reinforcement learning and programs like AlphaGo, which look at very important issues like what samples to learn from, how to decorrelate samples, how much to discount the future, how much confidence to give in one's current knowledge, and so on. I think there's a lot of potential lessons from what's learned there to complex systems in general, including economics. And I'd love to talk to people about it either at this panel or during the breaks. Thank you.

JOSHUA EPSTEIN We were invited to ponder the question, "What new idea, tool, or methodology should next be brought to bear to help close the gap between economics and complexity science?" In my view, economics needs to adopt a generative explanatory standard where we try to understand macroscopic social patterns and dynamics by growing them from the bottom up in agent-based models, artificial societies of cognitively plausible software people. It is of course an embarrassment that there are separate fields of micro- and macro-economics. Agent-based models are micro-to-macro models that can bridge that gap, explaining both macroscopic statistical regularities like scaling laws and so forth, and also nonlinear dynamics like financial contagions, panics, and market crashes. To do that, however, the models have to be populated, as I frequently say, with cognitively plausible agents. And, as has been amply demonstrated by Herb Simon and Dan Kahneman and the behavioral economists and psychologists and so forth, the canonical rational actor—a socially isolated, perfectly informed, computationally boundless utility maximizer—is not a cognitively plausible model of humans.

Rob and I always enjoyed it when we were economists at Brookings, our colleague Henry Aaron used to say that the rational actor is a passable definition of a sociopath.

[LAUGHTER]

So, unlike the rational actor, people have bounded rationality and systematically exhibit departures from the canonical picture as Blake and others have mentioned. There are many of these systematic departures, statistically erroneous heuristics, framing and conformity effects—a long list. But a long list of counterexamples and violations does not change scientific practice. As Samuelson purportedly said, "You can only beat a model with another model." Mountains of gripes and anomalies don't do that. We really need formal mathematical alternatives to the rational actor. I spent five years developing a very provisional candidate named Agent_Zero, who I think meets the minimal criteria of cognitive plausibility.

So, what are the minimal constituents of a cognitively plausible agent? I think there are three. One is that people have emotions. A plausible agent needs an emotional affective module—Agent_Zero has one of these. Adam Smith, whom we've quoted a lot, in his book, *The Theory of Moral Sentiments*, was really focused on emotions, and contagious ones at that, as was Keynes and his famous phrase "Animal Spirits," also the title of a recent book by Nobel Laureates George Akerlof and Robert Schiller. So, Agent_Zero can be driven by "animal spirits." Specifically, it has a non-deliberative affective module based on the neuroscience of fear using well-known simple equations developed by Rescorla and Wagner. Most social science is about choice. It's about decisions, it's about alternatives, and so forth. But we don't choose to fear. It's not necessarily conscious, and it's profoundly implicated in financial panics and other economic behaviors. So one component of a minimal cognitively plausible agent is emotions.

Second, yes, there's a conscious deliberative module, but it has to represent all the systematic—or some of the systematic—mistakes humans make in estimating probabilities, judging risks, even with perfect data, which we typically lack.

Third, these emotionally driven and statistically hobbled agents interact with one another in dynamic networks of their own construction, in which fears propagate, amplify, and generally produce highly dysfunctional collective behaviors. These include financial panics, but also genocide and the wittingly self-injurious health behaviors like addiction. Here, I don't believe in Gary Becker's rational addicts—his famous paper "A Theory of Rational Addiction" should really be called "An Irrational Addiction to Theory."

[LAUGHTER]

These endogenous networks, again, are based on the cognitive neuroscience of conformity. That's the third constituent of Agent_Zero. When

THE RIGHT IDEALIZATION

Joshua Epstein

Like many at the Santa Fe Institute, I am interested in making connections between physics and economics—but I am even more interested in *breaking* them! Economist Paul Samuelson saw the two as strict dynamical analogies. In his Nobel Lecture, we read, "If you look upon the monopolistic firm . . . as an example of a maximum system, you can connect up its structural relations with those that prevail for an entropy-maximizing thermodynamic system. Pressure and volume, and for that matter absolute temperature and entropy, have to each other the same conjugate or dualistic relation that the wage rate has to labor or the land rent has to acres of land." On the empirical fertility of this analogy, Samuelson also said, "Economists have much to be humble about."

Now, it is vulgar to criticize a theory because it is "wrong." George Box's quip is misunderstood. When he said, "All models are wrong, but some are useful," he wasn't saying that most models stink, but a handful still work. Rather, he was saying that all models are *idealizations* and, coyly, that some are extremely fruitful. No actual gases are ideal; no actual planets are point masses; no actual springs are perfect. But these idealizations, unobservable limiting cases, turn out to be the right idealizations for mechanics. I think the central issue before us is whether the "rational equilibrium picture," if you will, is the most fertile idealization for economics and the social sciences in general.

Beginning with Simon, Kahneman, and the entire behavioral economics "revolution," a mountain of fundamental departures from canonical rationality has been documented experimentally. It amazes me that the recent rise of behavioral economics should be considered "revolutionary," since it simply insists that we observe humans. It's as if, after 300 years of astronomy, someone had the temerity to suggest that we actually observe the planets! When we do observe them, that is, humans, we see that they do not behave like the rational actors of theory.

Rather, we are driven by contagious (and not always conscious) emotions like fear, what Keynes called "Animal Spirits." Our conscious deliberations, moreover, are hobbled by a variety of systematic errors: framing effects, base-rate neglect, anchoring, availability and confirmation biases, and conformity effects, to name a few. But anomalies and counter-examples alone do not change practice. The best economists will readily grant all of this but will say that, lacking a formal mathematical alternative, they'll stick with the rational actor, though they have in fact developed many epicycles (like the zoo of alternative utility functions) to salvage the picture.

If the rational actor really isn't the right idealization—if it's not the ideal gas, but just gas—then what is? I think we must develop formal mathematical and computational alternatives that are cognitively plausible, as licensed by behavioral neuroscience and related fields. In my view, the minimal constituents of a cognitively plausible agent are an affective or emotional "module," a bounded deliberative module, and a social module. My own simple provisional candidate, implemented first in differential equations and then in code, is *Agent_Zero* (2013). But, I am sure there are other candidates, and that machine learning, evolutionary computation, and other methods will help us discover them. Then, the data will decide.

Further Reading

Joshua M. Epstein. *Agent_Zero: Toward Neurocognitive Foundations for Generative Social Science.* Princeton University Press, 2013.

agents of this type interact, they do generate core collective dynamics and replicate some very well-known small-*n* psychology experiments.

My motto for generative social science is, *If you didn't grow it, you didn't explain it*. This is to be sharply distinguished from the obviously stupid converse that any way you grow it is explanatory. In addition, there may be many respectable ways to grow it, many cognitively plausible generators. In the Agent_Zero book and in subsequent work, I say the aim is not to get the modules finished, but to get the synthesis started. And in other publications, I have talked about alternatives to the specific affective, deliberative, and social modules used in Agent_Zero.

I think we need to discover families or neighborhoods of agents that are generative, not just single handcrafted isolas. Here machine learning and evolutionary computation can play a huge role in developing *inverse generative social science*, or iGSS. I like the acronym iGSS because it's (a) the inverse problem, pursuant to (b) the generative explanatory standard, applied to (c) social science specifically, hence iGSS. Here, we don't build the agents outright. We simply build libraries of rule-components and logical and mathematical concatenation operators and we turn evolutionary programming and other methods loose on those and grow the agents themselves. The fittest agents are those that generate the target data, whatever that might be—epidemics, revolutions, financial panics. The point is that we start with the target and evolve the agents. Agents become outputs, not inputs, standing the field on its head.

In conclusion, I would say economics needs to become generative. Agent-based models are the principal scientific instrument of generative social science. They have to be populated with cognitively plausible agents, grown with artificial intelligence and other machine methods. I think that's the future of economics.

D. C. KRAKAUER I want to ask some very big questions first, which haven't come up. One is actually the nature of computation itself—it's interesting that no one told us what it is. And for people in this room, they may think, well, what is that? Is that a Turing machine? Is that Python code? Are you saying that C is the alternative to ODEs, or are you more general than that? You're saying, "Actually, no, we're talking about discrete dynamical systems of a certain kind and there's a theory of those that isn't a computer simulation." Could the panel address these questions? What do you actually mean by *computation,* beyond using computers to do simulations?

. . . economics needs to become generative. Agent-based models are the principal scientific instrument of generative social science. They have to be populated with cognitively plausible agents, grown with artificial intelligence and other machine methods. I think that's the future of economics.

J. EPSTEIN I mean recursive, rule-based systems. The idea is that we're not using continuous differential equations and so forth. We have agents that are algorithmic actors. They take in information, they have processing rules, they adopt actions based on these. And, as Jessica said, I think social systems are doing computations in that sense; actual computer programs and so on are representations of those computations. But the computations are performed by the system proper.

M. MITCHELL I'm kind of on the side of the discrete dynamical system—not always discrete, necessarily, because we can have continuous value computation, but discrete dynamical systems that process information. I think the interpretation of that processing is an important part of computation, that not every dynamical assistant we could say is computing something. There has to be an interpreter, and it can either be somebody outside the system or it can be within the system itself, and in an environment in which the processing matters for some reason—that there is some kind of fitness criteria. But that, I feel, is an intrinsic part of what computation must mean.

R. AXTELL David, if you're asking what the theoretical foundation is and what the theoretical foundation should be, I think that today it's fair to say that there are a variety of ideas out there on how it might work and we don't have a consensus. Just to give a couple examples, there is this idea of what's called *interactive computation*. For those of you who have not encountered it before, Peter Wegner demonstrates that systems of interacting automata can affect or do things that are at least as powerful as a Turing machine, if not more powerful.

There's a whole different angle. To follow up on some of the things that Josh said, there are a variety of ways to motivate agents that are not really kind of utilitarian. For example, to have agents follow modal logics in which they're going to do things that are the norm, that they're obligated or they should do things. This would involve writing down agent specifications that use KD45, a very specialized formalism for looking at basically deontic logics of what could or should be done instead of what was best to the individual. So I think that at the research frontier there are a variety of ways to proceed. We don't really understand what the best way to go is, but it's actually an active area. Barbara Grosz is in the audience—if you go to an agent-based or an autonomous agent conference these days, you'll find a lot of computer-science theorists working on some of these topics.

J. FLACK I would say that, for adaptive systems, we don't actually know what computation is yet. We certainly don't want to borrow too heavily from computer science and certainly not ideas like Turing, I think, for reasons we could get into. But at the most basic level I would say that there's an input–output map. There's an architecture that explains how you go from the input to the output. There are circuits that describe the causal relationships or physical interactions among the nodes in the system over which behavior flows. There's a notion of correctness, which brings us to fitness, the idea that the output has to be adaptive, where we want to understand how it becomes more adaptive. And there is, as Melanie described, a notion of *interpreter*, which we think about as having agents who can read global information, or partly global information, in the system and make decisions based on those readouts.

So there has to be computation, and, to be honest, I believe that principle actually applies to computer science, too. I just don't think it's thought about hard enough there. Maybe, Melanie, you have something to say about that? There's a great paper that Walter Fontana found by Clare Horsman that was published a couple of years ago that makes this point for physical systems, that you need to have an interpreter in order to make the distinction between something that's a computer versus something that's performing a computation.[3] In adaptive systems, what I'm interested in is how that computation happens and what the basic elements are. I'll just add one more thing to that. At least as far as I'm concerned, once we have a good understanding of the architecture and how you get a notion of correctness or a termination criterion—which we can talk about—through a separation of timescales, once you get a good understanding of those things, then I think we need to develop a formal language or an underlying formal model of computation that would be analogous to things we have in computer science. But I don't think we can go there first. And that has been to some extent the tendency in importing computer-science concepts like Turing to biology: to start with that underlying model of how the computation works, which I think is wrong.

B. LEBARON I won't step into too much computation theory, David, because we've got all these other people to talk to about it and they're much better than me, but there's a very practical side. Modern macroeconomics is a highly computational field. It's standard to macroeconomics, and it's also a very heterogeneous option. It feeds in massive amounts of heterogeneous data to fit their large-scale models. Usually, the computer—what's its job? To find the fixed point. I spend a lot of time distinguishing myself, but it's not because I use a computer. They actually use more computer time than I do.

I hit the barrier as a practical thing because I'm continually dealing with different weightings. My agents are basically weighted on—well, once you have these distributions changing through time in really interesting ways, you don't have some sort of normal distribution or any

3 Horsman, C., S. Stepney, R. C. Wagner, and V. Kendon. 2014. "When Does a Physical System Compute?" *Proceedings of the Royal Society A* 470 (2169): 20140182. https://doi.org/10.1098/rspa.2014.0182

distribution you can do much with. You're kind of wedded to doing simulation at that point. I don't think I know the way. I really have no way around that.

J. MILLER David wanted us to disagree with one another, but I don't want to break the trend of disagreeing with David. So, where I disagree: he introduced this morning with, we have physics, we have computation, we have biology. Those aren't as clear in my mind, especially to this question of computation. Work I started a long time ago, inspired by taking the same class Melanie did at Michigan, was to evolve little discrete input–output machines with automata so often used as a model of computation, but to evolve those using the genetic algorithm. I think computation is taking place at multiple levels. Little machines do some computation, but certainly the evolutionary system is doing computation and altering what happens underneath.

D. C. KRAKAUER I want to turn this to the economy now. In the field that I'm closer to, which is intelligent systems, there's been a long history of asking what it is we're studying. What is the brain? Everyone here knows the psychohydraulic theories and electrical theories. I remember going to meetings where the brain was a search engine. They're all legitimate. Brian showed us the machine yesterday and he said, "The economy is not a classical machine." It's actually what Schrödinger would call a *statistical machine*. Again, there's been some, to Jess's point, emphasis on methods you use that would be best suited to the economy. But I would like you to tell us if you've thought about what the economy *is*. In other words, what kind of species of adaptive variety is it, and how does that map onto the theory of computation? Everyone has to answer that one.

[LAUGHTER]

J. MILLER Well, I'll start, to stay out of trouble later. Even the words you used—for example, ecology, species, adaptation—I think, really inform a view of the economy. It's a very complicated ecology that has a lot of parts and technology. I mean, you have physical things evolving and moving; you have human dimensions and behavior; and then you have these interactions creating new worlds, like banking systems and so on. So I think it's very hard to come up with a definition of economics.

The simplest one is the science of choice making, but that really doesn't limit it much. And, you know, systems make choices all the time, in all kinds of different ways.

R. AXTELL I would say that the classical—at least, twentieth-century–view of the economy is that the economy is a resource-allocation mechanism. That sounds somewhat archaic today, given that we know so much more about the economy. But it does grasp a certain elemental truth that we have to allocate resources if we're going to survive. We have to have a relatively efficient way to do that. This is Lionel Robbins, 1932: "An Essay on the Nature and Significance of Economic Science." This is a kind of foundational document for economics of that era and then going forward. It turns out, of course, though, that when Robbins wrote that phrase—you know, "resource-allocation mechanism"— there was no modern idea of computation.

Skip ahead to the '70s and '80s, and Hervitch and then subsequent researchers come up with the idea of a mechanism-design process. This reaches great import within economics, and no prizes are awarded for it. Now, sometime in the mid '00s, Tom Sandholm and a student, Vince Conitzer, at Carnegie Mellon proved that, in general, to solve the mechanism-design problem is an NP-hard problem. So it's not going to be one which is easily done, but they get the much more sophisticated result, which was: imagine that God hands you the mechanism. Imagine that God tells you that the mechanism is for allocating resources. If you're an agent in the system, it is an NP-hard problem to figure out how to behave *vis-à-vis* the mechanism. So the mechanism-design paradigm is not a credible one computationally.

In some sense, when we phrase it in Lionel Robbins's way from the 1930s, pre-computation, and he says the economy is a resource-allocation mechanism problem, today we know that that's essentially not computable. You can't use that as a proper framing. So, I think, from the modern perspective of computation, we know that the undergraduate specification of what an economy is, is naive and we don't really have a replacement, to be honest, today.

M. MITCHELL Just to push back a little bit on that, there are a lot of processes in nature that are technically not computable that happen all the time.

R. AXTELL Fair enough, if the system is local, but the economy is global.

M. MITCHELL Well, in any finite system, I think it's hard to say that it's not computable.

R. AXTELL But saying it's NP-hard is . . .

M. MITCHELL It's a little different.

R. AXTELL Yeah.

I resist the phrase *the* economy. When you ask what is *an* economy, there have been many economies through history and, yes, they allocate resources. They also propagate and sustain terrible inequality.

M. MITCHELL I like the reference you made that everybody's trying to explain the brain in terms of these metaphors. You know, Gerald Edelman tried to explain the brain as an immune system. I remember a talk at SFI a few years ago where I think it was Steve Zucker from Yale who said the brain's an economy,[4] but it goes the other way, too. People talk about the economy as kind of a big global brain. It's interesting to think about it that way because I do think that the brain is also a resource-allocation device, and it decides how we allocate our attentional and physical and emotional resources and it keeps us alive. But we actually do this allocation very poorly in some ways. I'm wondering, you know, does the economy also have its own sort of cognitive biases for resource allocation?

4 Mitchell is referring to a March 10, 2011 SFI seminar Zucker gave on geometry and perception.

J. EPSTEIN I resist the phrase *the* economy. When you ask what is *an* economy, there have been many economies through history and, yes, they allocate resources. They also propagate and sustain terrible inequality. This economy—the economy we're talking about—also does that. So, yes, the price mechanism is allocatively efficient under very stringent conditions. But issues of justice and equality and mobility and all of these things are not addressed merely by price mechanisms and other allocative machinery of the market economy.

B. LEBARON: You asked for these sort of biological connections. I think two hard problems we've faced all along at SFI are: thinking about technology—I'm going to stay out of that. The other is the evolution of institutions. Brian and I once thought institutional economics was very tied to what we were trying to do here. It's more important than ever. Someone mentioned yesterday—and I forget who—the work of Elinor Ostrom, who deals with the development of where these sort of useful social institutions evolve from. We don't know that. That's inherently an ecological thing. Can we answer it with our computer models? I don't know. I wish we could, moreover, in the ecological space.

J. EPSTEIN Yeah. I would just say, in addition, that a thoroughgoing Marxian economist would say the economy is a set of social arrangements that propagates the class structure.

D. C. KRAKAUER That's good. I like that. It's funny, right, listening to this panel on computation and the economy in 2019 and no one has used the words "neural nets" or "deep learning."

M. MITCHELL We said AlphaGo.

D. C. KRAKAUER Yeah, someone said AlphaGo. It was kind of hidden in there somewhere, and it's wonderful, because we're kind of Pluto in the solar system of computation in that respect.

I do want to ask a question which has to do with transparency and understanding. This is a debate that many of us have had for many years, and that is one of the limitations of computational frameworks. This notion, that you understand it if you grow it, might not be justified. These models aren't transparent to human reason and that for some reason is important to us. I'm just wondering whether you could address this issue both in relation to agent-based models, but also in relation to neural networks, which are the dominant computational framework today.

J. EPSTEIN Well, I don't want to talk out of turn. I think there's a lot of confusion about this. I mean, Alpha_Zero can crush humans at chess. But it is completely un-illuminating about how humans play chess. After he beat Deep Blue with a brilliant queen sacrifice, Kasparov was asked, "How the heck did you come up with that move?" He said, "It smelled right."

[LAUGHTER]

So, to be sure, AI can beat humans and augment humans. But, that is a far cry from explaining them. Used in this way, these machines don't give any insight into how humans do the things we're interested in. In this respect, I think the Turing Test is completely irrelevant to the social sciences. Behind the screen might be a human soprano and a perfect digital recording of her. That they are indistinguishable to the human ear tells me nothing about how the human soprano generates sounds, by pushing air across her vocal chords. Likewise, replication of social output by black boxes does not identify the social generating mechanism, but if we are trying to explain, that's what we're after. The mere emulation of output does not identify mechanism.

B. LEBARON I've built models with neural nets—with neural net traders. Generation two of the markets that I built back in 2000 or so had traders interacting with neural nets. They were built off them. Having thousands of neural nets competing with each other in a market is a daunting tractability issue. As a matter of fact, it's probably why I sort of had moved on from that. We're going to have to explore the space of the parameters in this thing much better than we did before.

It's going to be tricky, though. Often when we're developing these kinds of open-ended tools for agents, there's a great philosophical question of *what are we doing?* We've got the very simple ones and then we've got these kinds of open ones, but, when systems move out of what they're good at, they're a disaster. And we have these fancy ones that are designed to handle anything, but they're also driven by certain learning mechanisms. Is it a GA that's on top of them? What is it? That sort of philosophically doesn't make them all that different from the others.

R. AXTELL Machine learning and a variety of other fancy data-analysis techniques are all the rage today in certain parts of economics and finance where there are very important prediction problems for which we really would be happy to have an explanation for *why* things happen. But at the moment we just want to know—*will* they happen? The famous quip by Samuelson was that we predicted eight out of the last three recessions.

[LAUGHTER]

Just being able to forecast whether a recession is going to happen is something that's very important. Now, of course, there are these issues about whether having a credible forecast can modify the probability of it happening, but that's a different issue.

> . . . to be sure, AI can beat humans and augment humans. But that is a far cry from explaining them.

Anyway, there are certain questions in economics for which we would like to have better prediction ability, but I think it remains the case, at least from the point of view of the science of what would now be called *positive economics*. That is, why does the system do what it's doing? We really need to have acclimations that are couched in the behavior of individual agents and their actions. And so it may be the case that many of the modern deep learning and other kinds of things are not that useful once you get beyond the pure realm of prediction.

J. FLACK I have three related points in response to that. This discussion we have around neural nets being incomprehensible is a little ridiculous in a way, given the fact that this is typical of all of science. And so my three points are, consider how your brain makes decisions or how it performs computations when you're trail running and you're avoiding obstacles and so forth. You have no access to those computations. You can't articulate them. Maybe as a scientist, you could work out a model that would explain what's going on, but you can't relay that to anyone in this room. Right? So the way our brains work is inaccessible to us, just as the neural net is.

The second point is that most of the models we build in science—a lot of them—are incomprehensible. For instance, if you think about a metabolic network, you've seen in many people's talks here over the years, images of these crazy metabolic networks that have tons of edges and nodes and they're just a mess. But they're really rigorously derived from years of experimentation. And what we do as scientists—this dimension reduction to find out what the dominant causal features are—becomes our theory of how the output is produced from the microscopic. So we need to do something like that for neural nets through compression and other mechanisms.

And then the final point is that, of course, nature has this problem, too. Again, it's the collective computation perspective. Nature's in the business of building effective theories for how the world works, of simplifying the environmental complexity and making predictions forever. Sometimes it gets it right and sometimes it doesn't. But we can learn a lot about how to do that by studying how nature makes simple models of regularities.

J. MILLER Two points, one on neural nets: You know, we think of them as this recent innovation. I just want to point out the first neural net paper was 1943, so it's been almost eighty years for that development. On the transparency question, I think it's a good one. I agree with Josh—you know, if we can grow it to show it, great. But I also think there's value even if it isn't transparent. This came home to me when a couple of Go masters were talking about the AlphaGo game. It's as if an alien intelligence came down to the planet and we got to watch Go played in a way we'd never seen before. And that has tremendous value when you think about trying to understand the rules of Go and how one does well or not in these kinds of games. And I think, scientifically, even if things aren't transparent, but they're surprising, we may gain quite a bit.

R. AXTELL One other technical point about deep learners is, keep in mind that these are basically exotic regression models. These are in essence cascaded logistic regressions with many interior dimensions. And so it's certainly going to be the case that if you're going to be entering a novel regime for which the past data do not apply, the deep learner will go off the rails just as easily as in your regression model.

AUDIENCE MEMBER I saw Josh sort of laying down the gauntlet of what you thought the future of economics was going to be and that we needed to do a better job of these agents. And the practical thing that came to my mind is, why won't SFI put its voice behind, say, the development of an economic agent open-source platform? If you think about what Apple does with the App Store—you've got a million people out there using the platform. With that sort of practical approach, then we've got everyone working on what those agents are, refining them, where the fitness is, how it works in the real world. And then, in terms of deep learning, that's, in my mind, fine-tuning the parameters of these agents. But think of that as an ecology that we can use to build better models.

J. EPSTEIN Yeah, exactly right. I agree. I think that's an important initiative we should do. And I called it inverse generative social science, but in less fancy words it's exactly what you're saying. We have these economic dynamics and patterns and we believe that to understand you'd like to throw them in agent-based models, but we need to develop a big library of alternatives to the rational actor and explore their generative power. I think it would be a great enterprise for SFI to underwrite an advance.

R. AXTELL A different version of that might be the idea that we don't have in economics a set of community models the way that weather forecasters do. When weather forecasting, the Geophysical Fluid Dynamics Laboratory (GFDL) and Princeton and the National Center for Atmospheric Research (NCAR) and Colorado keep track of a variety of models for modeling the climate, for modeling various aspects of the weather, etc. And those models have evolved over time. They've become much more successful. It takes a hundred million bucks of annual funding to keep those things up and running. So it's a large investment that one makes in GFDL and NCAR. But it would be useful in the long run if we really imagine that you can have things like emergent macroeconomics, as Eric mentioned yesterday, or models in which macroscopic structure emerges from the microscopic actions.

You really need to have a set of community models. That is missing today in the economics and finance world. And it might be that an institution like SFI could take a role in catalyzing or fomenting such an undertaking—maybe not running the whole thing, but at least getting it off the ground and getting it started.

J. EPSTEIN Exactly. A public library of formal alternatives to the rational actor.

ERIC D. BEINHOCKER Just a bit of advertising that our group at Oxford has made a very small, modest start on that. There's a site on Github, *econlib*, for library, that is starting to build a collection of code in models in that spirit. But I think you're exactly right. It needs to be a kind of big, open-source, community-wide effort. And SFI should certainly be a central node in that.

I want to connect something that John and Jessica said in their opening remarks. John noted this huge kind of zoo of behavioral anomalies and asked what might provide some unifying perspective on that. And I think collective computation actually has the potential to do that. There's been some interesting work by cognitive scientists Hugo Mercier and Dan Sperber, who look at that library of anomalies and say, as an individual actor making an irrational choice framework, they make no sense. They're inefficient, they're biases. But when you look at it from a collective-computational perspective, a lot of those behaviors actually serve a purpose and make sense. They're features, not bugs. Now, some of them may also be bugs, but then, to Jessica's point, in collective computation you can weed out a lot of the noise and overcome that. So I think there's real potential to combine the collective-computation perspective with a kind of cognitive-behavioral perspective to make sense of the economy as an evolving collective computational system.

J. FLACK Just as a response to that, one thing that helps me clarify some of these issues with respect to biases and so forth is to think operationally about what *error* is. We talked about this last night. So, I define *error* as the deviation from the maximum theoretical structure that could be extracted from the environment, and keeping that clean. So that's my definition—or is that deviation? And then the separate question is function, right? So we separate this question of optimality into error and adaptive, essentially.

J. MILLER Yeah. I didn't want to be too hard on the behavioral economists, but if you look at this zoo . . . first off, the scientific incentives are horrible here where, if you can name your own effect, you get a lot of credit. And that's rolled into the experimental procedures, which are also pretty bad. And if you look at the 190 heuristics and biases, many of them are subtly different from one another, and also there's sort of this Newtonian idea that for every bias you can find an equal and opposite bias somewhere else.

[LAUGHTER]

J. EPSTEIN Very economistic notion, right? In equilibrium there's no bias.

J. MILLER On average, we're doing great, but I think there could be good starts of just trying to classify these and then think about error and try to unify them. So I am hopeful for that one.

J. FLACK Just one more point on that. This reminds me of Geoffrey West's work. He's in the room somewhere. So, in his city scaling work, there is this very large number of variables at the macroscopic level that seemed to be important. But many of them are probably nominal or correlated with each other in some way, right? Geoffrey, you jump in if you want. And in the metabolic scaling work, you have mass and metabolic rate, and Geoffrey and his colleagues have shown they can be derived from the microscopic. And so we have some better sense that they're fundamental.

We have the same problem with these behavioral biases, I think. What are the microscopic, neural underpinnings of these things? And if we can derive them from the neural understanding, then we can improve. We'll have a much reduced set. We wouldn't have 165 or whatever it is. We might have, you know, five. And I don't hear that micro-to-macro discussion with respect to behavioral biases. I mean, it came up somewhere, maybe in Josh's point about the agents as outputs instead of inputs, right? They're outputs in the sense that, if you model that underlying neurophysiological stuff, that gives rise to the behavior as an output. Right? This is where we bring neuroscience back into this discussion.

J. EPSTEIN I agree. The subtitle of *Agent_Zero* is "neurocognitive foundations for generative social science." I also love your point about inaccessibility. I mean, part of the agent is an affective fear module that is certainly inaccessible and, worse yet, it can override the accessible deliberative component producing very-far-from-rational activity. We're not doomed by that. But we need to understand it.

C. MÓNICA CAPRA I have a question. So, I'm a behavioral economist and every time I look at a Wikipedia page with a list of anomalies, I get dizzy, too. From an economics perspective, it's true that we have all these biases and cognitive limitations, emotion. I actually do some research on emotions. But what is interesting about human societies is that we have kind of evolved to develop institutions that help us overcome these biases, perhaps. So we have endogenous institutions arising in an economy. You can think of an allocation mechanism as an institution so that you can have efficient allocation.

So I was wondering, how will you compute these things? You have agents that are diverse, heterogeneous, they have all these biases, maybe informed from neuroscience, which I think is a really useful thing to do. But then what's happening is that institutions arise endogenously. Will the agents actually build institutions to kind of overcome these biases?

B. LEBARON I'll take a shot. I totally agree with you. I think that where we get effective institutions is very interesting. One big area in agent-based stuff, at least in the finance side, is what are known as zero-intelligence (ZI) agents. Gode and Sunder actually developed them and they get a lot out of looking at laboratory experiments with all of these complicated things. Pretty useful things emerged in the simulated market settings they were doing and we felt like it must be some big important human thing that we're doing. We're getting rid of the cognitive stuff and something good is happening.

Turned out they didn't make computer algorithms that actually had no intelligence, just like a basically falling budget constrains the mechanism of the institution. And, lo and behold, those systems were just as efficient as the humans. So, you know, it set a benchmark for us, defining this notion of how important the institution is. Are institutions sometimes evolving that help alleviate these biases or are there institutions that magnify them? We don't know. I won't go into that, but it's a super-important part. For us in agent-based modeling, at least in finance, the ZI—known as ZI traders, if you're cool.

R. AXTELL Quick comment on ZI traders. I was a graduate student when that work was being done. And the original idea was they had data on how MBAs actually traded in the lab and they had very complicated AIs as the MBAs, what they were doing, etc. And eventually, they found that they could turn all the complicated stuff off and explain the MBA behavior by just the zero-intelligence model.

[LAUGHTER]

J. MILLER Don't raise your hand if you have an MBA. Be cool.

R. AXTELL On the institutional part, I actually feel that that may be in fact the killer app for the whole agent-based paradigm in the sense that if you think, what will people be doing with computation in the year 2050? As we have so much more computing available, then the idea of having there emerge institutions of self-governance, of exchange, etc., and looking at alternative formulations of those things—that will really be something that will be a focus, where today we have very little of that work going on. I mean, just as a concrete example, the Acemoglu and Robinson work of about a decade ago was very popular as kind of a game-theoretic interpretation of how modern democracies arose.

And today I'm only familiar with one model that actually tries to grow that overall structure. My sense is that's a very pregnant area for more work.

M. MITCHELL Would you say that the internet started out as a distributed system of self-governance and kind of evolved into what it is today?

R. AXTELL Yeah, but at the microscopic level the ETM was running everything. The ETM was not evolved. The ETM was engineered; everything else has evolved.

M. MITCHELL Yeah, and it definitely has its institutions that both alleviate and magnify bias.

DARIO VILLANI Just a couple of comments. One is, this is an economics and complex-systems conference, not a computer-vision one. I'm amazed that every time people talk about neural nets, they talk about deep learning. If you have done any neural-net research on the *Financial Times* series, deep learning actually doesn't work. Everybody in the room would agree there is very low signal to noise. That means after a layer or two you have extracted everything. If anything, we should talk only about shallow learning. That's one.

Two, it's not true that if there is a world coming forward that is very different from the world of the past, the deep learning system or shallow learning system gets off the rail. The adaptiveness of machine learning is that over time you can transition across qualitatively different models. That's why that's the only shot you have at dealing with non-stationality.

And the third comment, you know, that's typical of most conferences I go to. People always say neural nets are just an old thing, actually discovered in the '40s, in the '50s. It has a very different form nowadays. You could also liquidate, in a cynical way, quantum mechanics, saying, what's so new? We knew there were particles. The only news now is that we don't even know where they are.

[LAUGHTER]

Just stating that somebody designed some neural net with activation in the '50s doesn't even begin to describe the revolution that is happening right now with neural nets and machine learning in general. So I think quoting the date just does a disservice to the revolution that is happening. Thank you.

TIM HODGSON[5] I'm kind of interested in computation towards what, or resource allocation against what, objective function. It seems to me that the economic machine that we have allowed to emerge is putting carbon into the atmosphere and plastics into the ocean and is employing children and perpetuating modern slavery, so how good is our resource

5 SFI ACtioN member and Co-Head of the Thinking Ahead Group, Thinking Ahead Institute.

allocation? How good is the aggregate computation and what levers can SFI science give us to adapt the machine?

J. FLACK I want to make one sort of point in response to that question, which might broaden the discussion a little bit, and that is, the economic system or any other system can be computing a particular output, like the distribution of inequality, so it has a particular form, or it can be computing something like robustness. These, I think, are two very different things. And so I think your question really is about the first kind of thing, but it might be that we want to be optimizing on the second.

> It seems to me that the economic machine that we have allowed to emerge is putting carbon into the atmosphere and plastics into the ocean and is employing children and perpetuating modern slavery, so how good is our resource allocation?

R. AXTELL I read the question as a policy one that actually asks how we get to alternative policies which *don't* pollute the ocean, etc., etc. And there I think that the obvious and very important contribution of SFI-type science is, simply, it is much less expensive to experiment with artificial systems and see what policies are likely to work than to try a policy in reality, have it not work, and then end up with an autocrat as president, for example.

[LAUGHTER]

So, I think one very important role that we can use these technologies for is for policy simulators, broadly construed.

BARBARA GROSZ[6] I want to pick up on this last discussion and also ask you to comment on the connection between this panel and the first session yesterday. My read of the first session yesterday was that it really

6 Former SFI Faculty and Science Steering Committee member; Higgins Research Professor of Natural Science, Harvard University.

spoke to us about how to change the economic model, to deal with some of the problems that are here, and what the political sphere was. Much of the discussion in this panel is how to model how things are now, except for this last comment. It seems to me that that difference is huge, and this difference relates to the question of interpretability. If you're just modeling an existing system, maybe you don't need to explain causal connections.

If you're trying to do what Rob just brought up, then it's essential to understand where the causal levers are, or you're in a situation of having a combinatorial explosion of the parameters you're going to change to see if you can have an effect. So, if you could relate the kinds of models you're talking about, the kinds of agent-based modeling you're talking about, to what the speakers in the first session were trying to argue for, I think that would be helpful.

J. EPSTEIN Well, I loved the first session and think Eric was talking about what I would call *computational humanism*. We'd like to use these models to understand the emergence of highly unequal systems at all levels. I mean, this business about Pareto laws and skewed distributions—those are characteristics that permeate economic life. Rob has a fundamental model on the emergence of the firm size distribution, but there are others on the emergence of skewed wealth distributions, very unequal opportunities for advance, educational systems, all these other things.

I think SFI could bring into view a much clearer picture of how those inequalities emerge and, thereby, how they might be disrupted, in particular by micro-interventions that change the emergent phenomena. Again, as Rob was saying, from a policy standpoint the agent-based model is a way of exploring how micro-interventions can produce macroscopic changes. Matt Jackson's work also bears on this.

M. MITCHELL Yeah, I think that the key words you used were "interpretability" and "causal," and I think that's something that SFI can get at by moving away from the current paradigm of deep learning, which doesn't feature causal models or interpretability. And I think that's a really important point that we have to keep in mind.

THE FUTURE OF COMPUTATIONAL ECONOMICS

Blake LeBaron

Computation and complexity have had a significant impact in finance, but much progress remains to be made in terms of general acceptance and use as tools for solving large-scale financial problems.

Key to the history of this area was the development of the Santa Fe Artificial Stock Market roughly thirty years ago. A team that included Brian Arthur, John Holland, Richard Palmer, Paul Taylor, and me constructed an agent-based financial market that was a proof of concept for early explorations in complexity economics. It demonstrated that a diverse set of purposeful but relatively simple computational agents could replicate many basic empirical features seen in most financial time series. These were often properties that traditional models were incapable of hitting, like fat-tailed return distributions, periods of high and low volatility, and large swings in price levels.

There was an eventual explosion of models able to generate most of these macro-level features emerging from the interaction of simple adaptive agents. This has left some confusion in the field, with too many models and parameters chasing too few high-level facts.

Over this thirty-year span, economics has changed by becoming more accepting of behavioral biases, something others on this panel comment on. This has also broadened the set of data that researchers are using, including individual agent forecast data that often reveal irrational behavioral patterns that line up with agent-based models, but not more traditional rational economic frameworks. Another less-than-rational behavior that is well documented in financial data is the tendency of investors to chase good performance; they move capital into mutual funds that have shown good performance in the recent past. This human behavioral feature is often part of the core design of an agent-based financial market as adaptive agents chase strategies that have done well in the recent past.

Where is the future in agent-based financial markets? Obviously, vastly improved computing power and better data availability are changing the way things are done. The data front may not be as radically diverse as in other fields of economics. Finance has always been awash in good data, and some of the really interesting information remains locked in proprietary vaults.

Agent-based computational models have remained surprisingly quiet about multi-asset market relationships. Empirical relations across assets and asset classes is somewhat foundational to all of finance, but remains only a small part of agent-based approaches. These areas remain open for new approaches that envision correlation patterns as endogenous emergent properties, as opposed to being driven by underlying economic fundamentals. This is also related to the question of "crowded trades," where trading strategies themselves are piled on by many agents driving correlated buying and selling, which can lead to price instabilities.

A final way current financial market commentary may have caught up with early agent-based models is viewing markets as an ecology with interesting changing populations. Recent debates about the increasing fraction of passive, index-tracking traders and their impact on market dynamics fit into this discussion. Computational agent-based approaches are ready to step into this debate since, at their core, they are built to study the population dynamics of diverse trading populations.

Agent-based tools have enabled researchers to view markets from a complexity perspective, providing a different way to think about the dynamics of human behavior as it pertains to trading in modern asset markets.

B. LEBARON I was just going to bring up an example of taking agents into a radically different situation. They're trained up in one situation. There was a project that came out of SFI—actually really done by a consulting firm named Eurobios, by a guy named Vince Darley. Many of you may remember this—they were attempting to understand what happened at NASDAQ as we moved from eighths to decimals. They basically built an agent-based system and the agent-based system was going to take it into something that had never been seen before. For example, in the big-picture climate stuff, you're going to take it out into an area that's never been seen before. The system proceeded to make a whole bunch of predictions—and they're actually in Vince Darley's book—which are really quite amazing.

One of them was the prediction of high-frequency traders, and these were not on the landscape at that time. So whether that's just an amazingly lucky, great example of things working well in terms of understanding policy outside the box, I don't know. There are various experimentalists who tell me that they are able to do the same thing in the experimental world. I think our ability to do this . . . maybe don't write it off yet.

R. AXTELL But then, more generally, it may be the case that every time there's a policy change, there's the possibility of unintended consequences. It's going to be very hard to get at those analytically, but with an open-ended computational platform, you have at least some chance of seeing them ahead of time—seeing them ahead of when they happen. Getting some head start on trying to ameliorate their effects.

J. FLACK So, Barbara, I agree with you. In fact, that's one of the motivations behind the collective-computation work—to make better micro–macro maps. We know how to do the intervention more precisely, and it brings up a number of interesting issues. One is, Josh was saying intervention at the microscopic scale, but it's not at all clear whether that's where the intervention should be. Maybe it should be at the mesoscopic scale using dimension reduction, as an example. This is a big debate in biology right now and I'll give you the sort of collective-behavior take on it. In the collective-behavior literature, there are some researchers who like to get the rules and interactions at the individual level and build models from the bottom up. And then there are others who use more of a fluid-dynamics approach and they build these mesoscopic models.

As an example, there was a recent paper published on predicting movement during marathons. These researchers used the fluid dynamics approach to model marathons and it was a very mesoscopic-level description and they were very effective at predicting. They used data from the Paris marathon and they were able to predict New York, or something like that. They contrasted this approach in the introduction to their article with the kind of microscopic approach other researchers are taking. I feel this is a very important distinction to put on the table, but that they misunderstood something critical, which is that both groups of people want to know different things, right?

So, what the fluid dynamics people learned is that they didn't need to know anything about individual behavior to make predictions about crowd motion. But we sometimes want to know how individuals, by changing their rules, can influence the output of the system. And that requires maybe more of the microscopic approach. But once you have that big, elaborate microscopic map, you can probably—hopefully—do some dimension reduction on it to get the most important causal features. So, the question is, do you need to go through the microscopic to get to the dimension reduction? Or can you start at the mesoscopic scale?

D. C. KRAKAUER: Thank you very much to all of the panelists.

PHYSICS & ECONOMIC SYSTEMS

[Panel Discussion]

Moderated by David C. Krakauer,
featuring Ole Peters, Maria del Rio-Chanona,
Cosma Shalizi & David Wolpert

DAVID C. KRAKAUER Early economic theory—the kind that Ken Arrow, who has been mentioned many times here, talked about—was extremely enamored of the physical perspective, not least, their search for these very low-dimensional explanatory measures.

Ole is a good example of that, actually. Ole is substituting one low-dimensional measure for another. It's a very physics-based perspective and he's even got a physics term in it, which is very different from the previous panel. I mean, you're saying, "Look, it's just not the ensemble average; it's the time average." If we get the measure right, we'll have the right variational framework for doing economics. That's the sense in which I mean they're different.

I would ask all the panelists to tell us who you are when you make your opening remarks.

COSMA SHALIZI I'm Cosma Shalizi. I was a physicist up through my PhD. I have gone native as a statistician. So, my physics heritage at this point consists of doing sloppy math because I think I know what the right answer should be.

[LAUGHTER]

Let me start it off by throwing a few bombs. Despite a lot of effort by a lot of very smart people, I don't think that physics has contributed all that much to our understanding of the economy. There are two reasons

OPPOSITE *Economic Curiosity: Virgin Mary Grilled Cheese Sandwich* for this. One is that the disciplinary training of physicists—at least, theoretical physicists, who are the ones who have tried to do this, because we feel like we can do anything—neglects actually grappling with data. And, when you try to deal with real-world systems and not what the experimentalists hand you as a canned description of an effect, then you have to actually be able to do statistical inference. You have to actually be able to do data analysis.

This is something physicists don't really get trained in. That's improved over, say, the last twenty years—maybe especially over the last ten years. But a lot of what we have done has been very crude curve-fitting and looking for crude patterns or throwing a curve-fitting at it and hoping to predict.

Despite a lot of effort by a lot of very smart people, I don't think that physics has contributed all that much to our understanding of the economy.

The other reason is that we have this impulse to look at something with a lot of interacting parts. We say this is a many-body problem. I did this in graduate school. I was trained to death in doing this in graduate school. How hard can it be? It's either going to be the Ising model or it's going to be hydrodynamics. I've done that problem. And the difficulty is that there are just many, many other kinds of bodies. The limited repertoire of models we learn for saying how different bodies interact and how you aggregate their local interactions into large-scale phenomena—that part of our impulse is completely correct, but the repertoire we have of how you actually do this is just ridiculously too limited to grapple with what we see in the economy or what we see in evolutionary biology or a whole bunch of these other fields where physicists have tried to take them over, and where there is a history of SFI-inflected attempts at disciplinary imperialism, which have been more or less successful.

So, I would say that the reason econophysics has not really contributed all that much to our understanding of the economy is that it hasn't thought enough about the economy and it hasn't been good enough at thinking about different sorts of bodies and how they might interact.

At some level, yes, it's a many-body problem. At some level, with enough bodies, with enough independent actions, there's going to be some sort of large-deviations principle. Maybe something vaguely like statistical mechanics will emerge out of that, but we shouldn't expect anything like what we are used to to come out of the kinds of economic systems we care about.

Now, this is probably not what many of us would like to hear. This might just be too much of a negative way to open things. So, let me just say, by way of even-handedness and negativity, the economists have their own problems.

[LAUGHTER]

I'm trying to remember the exact phrase that came up during one of their internal disputes, which is that theory is evidence, too.

When you have some apparent empirical result, which contradicts some cherished belief of the profession, then the impulse is to just say, well, that can't be right, and to then enshrine various bits of theory very deeply in the internal psyches of the profession and in the internal professional hierarchy of who gets to say things which anyone pays attention to, as opposed to going to the journals full of cranks.

And, you know, this shows up in their ideas about individual decision-making. It shows up in some of the ideas about institutions, and what is the ideal form of institution, and it certainly shows up in macroeconomics, an area where I've been working a lot. You can demonstrate things like, if you believed one of their baseline macro models was correct, and you simulated it, you would need about 1,500 years of data to estimate the parameters from the correct model.

The reaction to this is about what you could expect if you've ever talked to a macroeconomist. With that, I'm out of time, and so I will hope that other people will poke holes in everything I've just said.

MARIA DEL RIO-CHANONA Thank you, Cosma. I'm Maria. I'm a PhD student of Doyne's. I work at the Institute for New Economic Thinking (INET), where Eric Beinhocker is, as well. I'm going to start by talking about what I think physics can provide to economics, and then I'd love to reply to some of the things Cosma said.

So, I think an idea from physics that economics needs to take is this: we need to start observing the economy. We need to start observing how the economy evolves. Instead of thinking, "Huh, the economy *should* be at this state," how would it fluctuate around such a state? Because that's what equilibrium is.

We think, "Oh, the economy should be here," because if it wasn't there would be a law. There would be the invisible hand. They'll put it right back in there. And this is how the Greeks thought about physics. They thought about it in ideas. They thought heavier objects fell faster than lighter objects, and that was natural law.

Physics started making progress when we actually started seeing the world. And how do we see the world? Well, we see the world through data. I do believe that's something physicists have started doing. Now, there's a difference between physics and economics. You might say it's easier to see the physical world than the economy. Well, that might be true, but it's not that easy to see the quantum world.

Still, we were able to see it through experiments, and at CERN, there's a lot of data being gathered every day. In economics it's harder. But we also have one advantage. The advantage is that we are the micro scale and we have some intuition about how it works, which is different than in physics, right?

I haven't heard anyone say, "Oh, quantum physics—that's so intuitive. It's like how I went to school." Right? It's not like that, and we do have some intuition about economics. I think right now the person who is not able to analyze data and find hidden patterns is blind. That person cannot see the world.

Because right now, with the amount of data volatility, we can see many parts of the world we couldn't see before. And this has to change not only in economics, but in the way we educate our system. Because every person who doesn't know how to use data—we can't hope for them to understand the world. Not this world.

COMPLEXITY ECONOMICS FOR A PANDEMIC

R. Maria del Rio-Chanona

Several years from now, if people ask me to define *complexity economics*, I'll tell them about the COVID-19 pandemic, SFI's call for a New Complexity Economics symposium three months before, and the Institute for New Economic Thinking at Oxford Martin School (INET)'s answer to this call. I'll explain that, during the pandemic, many people were unable to work, some goods stopped being produced, and it was clear that the supply of goods and labor did not equal their demand. Governments struggled to stop the spread of the epidemic and mitigate the economic damage. Most economic models were built to analyze an economy in equilibrium (or close to it), in which supply equals demand. I'll tell them that, three months before the pandemic started, the SFI community had discussed the need for out-of-equilibrium[1] economic models, and the role data played.

At the start of the pandemic, my INET colleagues and I began to build one of these models. We collected data to understand supply shocks, such as labor shortages in non-essential industries, and demand shocks, such as a decrease in the number of people using public transport.[2] Since industries rely on other industries for inputs, these shocks propagate through the economy; essential industries may be immune to direct shocks, but will be affected when shocks hit their suppliers. We modeled shock propagation to determine the overall economic impact.

The model's complex dynamics reveal counter-intuitive results.[3] For example, re-opening only a few industries could decrease economic output. This happens when supplier industries cannot produce as much as demanded, and other industries compete for these scarce inputs. When more industries open, the competition for scarce goods increases. As inventories run out, important industries may lack inputs critical to operation. Neglecting the interconnected nature of the economy or assuming a rapid adjustment to equilibrium yields different results and policy recommendations.

Most economic models were built to analyze an economy in equilibrium (or close to it), in which supply equals demand.

Data and heterogeneity between countries also play an important role. Few workers in developing countries can work from home and are thus suffering significant economic losses

1 I refer to equilibrium models as those that assume that supply equals demand, while out-of-equilibrium models do not enforce this constraint.

2 del Rio-Chanona, R. M., et al. 2020. "Supply and demand shocks in the COVID-19 pandemic: An industry and occupation perspective." *arXiv preprint arXiv:2004.06759*

3 Pichler, A., et al. 2020. "Production networks and epidemic spreading: How to restart the UK economy?" *arXiv preprint arXiv:2005.10585*

(Djankov and Panizza, 2020).[4] At INET, we are strong advocates of data-driven models and are currently recompiling data for several countries to test our model. However, few developing countries have the high-quality data needed for calibration and model testing. This means that we should collect better data ourselves—or build models that require less of them. I am not very optimistic about the latter's plausibility, but I cannot deny that models with minimal data requirements would be immensely valuable.

Had more people been working on data-driven and out-of-equilibrium economic models before the pandemic, we would have been better prepared to face it. My takeaway from SFI's meeting in November and recent events is that:

i) Complex-systems researchers should actively seek to interact with the economics community and help establish out-of-equilibrium models. To do so, we should read the economics literature and publish in economic venues. We must remain critical but also acknowledge there is work of excellent quality in economics.

ii) We must think of creative ways to collect data, test our models, and provide alternatives when data are scarce. Next time there is a crisis, we must be better prepared.

4 Djankov, S. and U. Panizza, *COVID-19 in Developing Economies, A VoxEU.org Book*, CEPR Press, June 2020.

So, why do I think physicists are trained in that way? Maybe it's because I did my undergrad in 2012. My first course was on programming and I did a bunch of experiments. Now, my undergrad was back in Mexico. The lab stuff we had wasn't very advanced, so half of my lab reports were about explaining why the experiment didn't work. That's why I thought I might have a shot at the economy.

[LAUGHTER]

So, I agree when you say we shouldn't categorize the economy. We shouldn't observe the economy and be like, "Oh, that's a harmonic oscillator." Yeah. "That's an Ising model." No. But we *should* absorb the data. We *should* observe the economy, and think in the physicist way of saying, "Okay, this is what I see. Let me think of a way of describing it." That, for me, is the only way we can actually observe stuff and stop being blind. So, I'll leave it at that.

OLE PETERS Great. That means I'm up. Cosma said that physicists are no good at economics because they don't look at data, and then Maria said that's not true; physicists are great at economics because they do look at data. So I'm going to take the position that we're great at doing economics because we don't look at data. This will lead David to argue that we are terrible at economics because we look at data.

[LAUGHTER]

The serious point of this is . . . it is true that, in the work that I've been doing, we have started from really simple models. We've done pure theory, and then taken the theory to the data, and the data sometimes told us that we missed something. And so we had to adjust the theory. But, in general, the kind of paradigmatic way of doing science as one of the routes has been working quite well for us.

So, sit and think and try to figure out what might make sense in what is a very, very simple model—think about what kinds of effects you may be able to explain with such a model, and what is beyond that model, and then go and take it to the data and see what happens.

Newton has come up a couple of times in these discussions. I just wanted to say I don't really think of Newton's equations as an optimization problem, but, rather, as dynamics. They're differential equations.

I think of them actually more in the computational sense. And maybe this is how I think about a lot of physics: that you're building models by writing down differential equations, which are updating rules. It's not so far from the computational approach, perhaps.

So, are there Newton's laws of economics? Well, yes, of course, but they are just Newton's laws. If you take an economics textbook and do something useful with it and throw it out the window, it will follow a parabola just as Newton's laws tell you.

[LAUGHTER]

But aside from that, I think there is something in economics—at least, to some extent—that's sort of the wonderful thing about Newton's laws, and that is universality. There is some degree of universality in economics and I just want to mention one example.

It's an equation that we now call *reallocating geometric Brownian motion*. Geometric Brownian motion is just a stochastic differential equation that describes multiplicative growth in continuous time. It's one of the standard models for asset-price movements in all sorts of things.

Physicists noticed one thing about this equation. It's the following: If you have a large population of people and they all have some wealth, and this wealth is described by that equation, geometric Brownian motion, and you wait for a while, you observe something they call *wealth condensation*.

Essentially, one person ends up with all the wealth in the system. This is just what that system naturally evolves to; it's just a mathematical result. That sounds a bit unrealistic. You wouldn't really think that this is how our actual personal wealth behaves if we track it all the time, because we don't live in a world where one person owns everything.

Now, timescales are very important, because it will take a long time. But in principle we'll start adding correction terms to this very, very simple model. And the simplest correction term I could think of was just a coupling, where you'd say, "Okay, we're not independent. We all somehow are coupled economically to each other."

Let's just say we all chip in some sort of contribution that we make to society, and we receive something out of this pool that we create by pooling some resources. This creates this equation, reallocating geometric Brownian motion. And we start studying this equation and realize that it has some very interesting features. It generates a realistic wealth distribution with a realistic power-law tail.

You could study these transitions from ergodic to nonergodic behavior. It's very nice. And this is where the universality comes in. We realized that we weren't the first to discover this equation. In fact, there were four separate groups of physicists who had all converged on studying this same identical equation.

> So, are there Newton's laws of economics? Well, yes, of course, but they are just Newton's laws. If you take an economics textbook and do something useful with it and throw it out the window, it will follow a parabola just as Newton's laws tell you.

They all converged on that somewhere in the last maybe fifteen years or so. The reason for that is that the mathematics to analyze this equation was developed in the 1980s and 1990s. This is really recent stuff. We all converged on this sort of dynamic simple model and computational, dynamical, differential equations. It's quite a powerful model in the realm where you would expect it to be able to say something. And I'm out of time. So, with that, I will hand it over.

DAVID WOLPERT I want to say something about the oversights, the lacunae—I think maybe Rob Axtell used that word—in our current thinking about economics.

There is a particular economics phenomenon which is almost definitionally far beyond physics or any known kind of modeling, and which I think has the possibility of occurring quite soon. I'm not going to say this phenomenon will definitely occur. But there's a non-negligible probability that it will occur within about a decade or so, and if it does, and if sometime afterward we were to reconvene and look back at our conversation today, we would think it's all quite quaint and cute and the little hiccups and burps of children.

This phenomenon I'm describing would arise due to three technologies that are currently converging. None of them have fully developed yet. But all three of them might develop in such a way that very interesting phenomena would result.

One of these technologies is blockchain. What's interesting about blockchain is not cryptocurrency. That's noise, ripples on the surface of the human economy. To be quite honest, it's not particularly interesting from a scientific point of view, and it's not going to have a major effect on humanity.

What's interesting about blockchain instead is that it enables "frictionless binding contracts," to use the language of economics. What that means is a bunch of things. One of them is that we might now be able to finally follow along with Willie Shakespeare and get rid of all the lawyers, since there would be no more need to have them enforce our contracts. More generally what it means is that all kinds of different contractual relationships that right now cannot come to pass because you have to have lawyers involved either to set up the relationship or to adjudicate it—which, also, by the way, contributes a lot to actual inequality because the legal system, of course, is not equal—all of those frictions due to the need for lawyers will be removed so there will be many more contracts.

What's interesting about blockchain is that it enables "frictionless binding contracts," to use the language of economics.

Okay, fine. Good. That's interesting. But there's something even more important about this. The so-to-speak default mode of blockchain contracts is that they are publicly visible. The reason that's important is that, as soon as you have publicly visible contracts whose payouts are contingent on events in the world, like what the people party to the contract do, that means in many, many such situations—this has

THE (NOT) SHORTCOMINGS OF MODERN GAME THEORY

David Wolpert

Modern game theory is the most broadly and thoroughly researched mathematical framework we have for analyzing strategic interactions among agents with conflicting goals.[1] Nonetheless, it is fashionable these days to pick on economics in general, and noncooperative game theory in particular. Unfortunately, this is often done without first going to the trouble of learning what modern game theory actually says.

As an example, skimming the literature, one quickly learns that the bounded rationality of real humans has been a major focus of both theoretical and experimental game theory research for decades.[2] Yet it is fashionable among those who do not know this literature to claim that game theory assumes humans are fully rational, and to dismiss it accordingly. Such attacks on modern game theory based on the flaws in *Homo economicus* are akin to attacks on modern biology based on the flaws in Lamarckism. Similarly, the consequences of multiplicative rather than additive payoffs have been central to finance theory for a very long time; to ignore that earlier research while investigating multiplicative payoffs is inefficient, to use a euphemism.

Nonetheless, once one becomes conversant with the noncooperative game theory literature, one realizes that there are indeed shortcomings in its very foundations—shortcomings that prevent it from making predictions about the vast majority of non-laboratory, real-world scenarios. It's just that these limitations are not the caricatures described above. Rather, they are far more challenging to address (which is part of why game theory has not addressed them!). One of these shortcomings on which I have worked is the specification of the move spaces of the players—specification in one's model of what actions each of the interacting humans might take.

In the vast majority of real-world interactions of humans, the sets of possible moves cannot be specified ahead of time. For example, even enumerating the possible statements people might make to one another in their conversations "in the field" (rather than in a controlled laboratory setting) is not possible. To enumerate the set of possible resulting actions those people might take is simply not conceivable.

Such attacks on modern game theory based on the flaws in *Homo economicus* are akin to attacks on modern biology based on the flaws in Lamarckism.

Unfortunately, noncooperative game theory requires complete specification of the possible

1 See Fudenberg, D. and J. Tirole. 1991. *Game Theory.* Cambridge, MA: MIT Press; and Osborne, M. and A. Rubinstein. 1994. *A Course in Game Theory.* Cambridge, MA: MIT Press.

2 Camerer, C. 2003. *Behavioral Game Theory: Experiments in Strategic Interaction.* Princeton University Press.

move spaces of the agents just to define the associated strategic scenario, never mind make a prediction of the joint behavior of the agents in that scenario. This means it cannot be applied to the vast majority of real-world scenarios.

Okay, but how could one possibly formalize an interaction among strategic agents without specifying their possible move spaces? As it turns out, the second major type of game theory—cooperative game theory—provides one answer. In general, in cooperative game theory one only needs to specify the joint utility values that the interacting agents assign to the possible outcomes of their interaction, and does not need to specify the sets of moves the agents can choose.

Unfortunately, in its standard form, cooperative game theory can be used only to predict the outcomes of bargaining scenarios. This focus on bargaining is hard-wired into the foundations of cooperative game theory via an axiom of Pareto efficiency.[3] However, it turns out that one can remove the Pareto efficiency axiom in cooperative game theory and still have a self-consistent, powerful formalism for making predictions about the joint behavior of the agents. Moreover, lacking the Pareto efficiency axiom, this hybrid formalism can be used to make predictions about noncooperative game-theoretic scenarios even though one never specifies the move spaces of the agents, in contrast to standard noncooperative game theory (Wolpert and Bono, 2013).[4] As an added bonus, bounded rationality is built into the foundations of this hybrid formalism.

How useful/accurate is this formalism? TBD. The important point does not involve that particular formalism, though. It's rather that the shortcomings of noncooperative game theory are not the ones that are so popular to pick on—there are others, far more profound, and research should be directed at those shortcomings.

---

3 Per Oxford Reference, *Pareto efficiency* is defined as an economic theory in which "an alteration in the allocation of resources is said to be Pareto efficient when it leaves at least one person better off and nobody worse off."

4 Wolpert, D. and J. Bono. 2013. "Predicting Behavior in Unstructured Bargaining with a Probability Distribution." *Journal of Artificial Intelligence Research* 46: 579–605.

been proven mathematically—a third party can come in, offer new contracts to the people involved in the original contractual relationship and completely screw up that original contract. This is all going to be legal. Completely screw up the contract in such a way that you're going to change the outcomes of that original contract in an arbitrary way. The third party coming in and doing this to the people in the original contract will be able to treat them and their contract as a money pump.

You can just pump money out like this in many, many situations. This kind of money-pumping might in fact already be going on in Wall Street. We don't know. It would be hidden. We want to actually make this visible. But the mathematics is very, very simple, non-cooperative, single-stage game theory. That by itself starts to become interesting once these kinds of contracts get deployed out in the economy as a whole. So it's much bigger than the financial system.

Things become yet more interesting when you start to realize that, if I tried to money-pump one contract, somebody else can come in and try to money-pump me. You get an entire eruption. Basically, there could be an ecosystem of these things interacting and we don't know how to model it. The problem is, you're talking about the space of all possible contracts. That's Turing complete. We don't know how to get our mathematical handles around this. Please understand, I'm not just spouting science fiction. I'm not just talking about technical gobbledygook of different modeling choices. There's a good probability that these things will happen.

I can't give you any more precise focus, a vision of what this could be, because it's completely unforeseeable right now to us with the tools that we have. With any kind of modeling tools that we have.

That was the first technology. Contracts, buy any contracts and they go by blockchain. The second one that was mentioned before is AlphaGo. Many people already are very comfortable basically giving automated agents access to their bank accounts and permission to act on their behalf—flash trading and so on. Take those agents and give them some of the capabilities of AlphaGo, make them reinforcement-learning algorithms, especially some recent results in symbolic regression, thus making this far more likely to occur: let them loose, going out and, on your behalf, trying to find contracts with the ones that are publicly visible, or even ones that you just suspect are money pumping the system, while other such agents, just like AlphaGo, coming up with strategies that are beyond human ken, are going out and playing the exact same thing. We don't know how to regulate this because we have no idea of what the actual space is.

All of this is likely. I'm not going to say it definitely has to happen; there's still advances to technology needed. But all of this is likely to be occurring within five to ten years. And, if it does, it'll make all the current talk about the singularity in terms of a single artificial general intelligence (AGI) becoming more intelligent than us seem particularly small-minded. That the overthrow, so to speak, of our current economic system—in other words, our current ways of interacting with one another—would occur by having a distributed ecosystem of agents that are interacting in terms of binding contracts. That could happen within the next five to ten years. It's out there already and we have no idea how to even think about it, never mind try to model it, never mind try to get ahead of the curve. So, way beyond physics.

D. C. KRAKAUER This leads to larger questions. Bearing on the first two remarks, the first is really that we inherit a set of basis functions and models that we use as lenses to look at the world. And there's no doubt about the allure of simple dynamical systems, the most obvious coming from ecology—Lotka–Volterra-like thinking. I've seen it in every domain from economics, biology, Ising models, variations of those models.

Physics has been very successful at breaking out of the shackles of its own bases, quantum mechanics being perhaps best known. Is there in that domain some evidence or some suggestion that physical frameworks themselves are being challenged from within the field? I know that people like Rovelli—and we might not agree with what he's saying—are arguing for alternative physical frameworks. But is there some suggestion that, perhaps along the lines of Ole, it was more of a mathematical innovation? There might be new potentially useful physical theories that we're not familiar with.

C. SHALIZI I'll bite. There's a lot of work on active matter. And there's a lot of work on self-propelled particles. This would be at least a step towards the kinds of things we're interested in. But that's no evolutionary development within physics.

I don't think there is any particular reason to think that new development of physical theories for physical problems, whether it's gravitation or what on earth is happening in a sand pile—*and* can you actually figure out where the forces are—will have any relevance to economic modeling.

If we are interested in economic modeling, I would humbly suggest we should think about the economy, and do some empirical work on economics and into the decision making in institutions. Some of this will involve, unfortunately, actually listening patiently to economists when they point out that they've had models of wealth distribution based on multiplicative growth since the 1950s. They've been discrete-time models. They've not been quite as nice as coupled stochastic differential equations. But if you go back to work in the '50s by people like Champernowne, by people like Herb Simon, by a bunch of others whose names are eluding me but Matt Jackson could probably tell us, they worked on models of multiplicative growth of wealth and showing how this led to at least not crazily unrealistic distributions of wealth over the population.

And then you have questions about how you can ground those phenomenological stochastic-process models and actual economic processes of distribution or exchange or production. But that would be a starting point of saying, "Okay, people have been thinking about this." And there are some phenomena which we know those have a hard time explaining. You know, there are destitute people, but there are limits on destitution, and so on. There is stuff to learn from them. Some of it is very basic stochastic-process stuff, which looks like basic stochastic-process stuff in physics or biology or whatever else because it's very basic mathematics. But that is the kind of learning I think we're going to have to do more of.

One thing I will say is that there are places where this is going to intersect with the computational stuff we talked about before. Zero-intelligence agents came up in the previous session. There is a wonderful paper by Gary Becker about the foundations of equilibrium analysis.[1]

---

1 Becker, G. S. 1962. "Irrational Behavior and Economic Theory." *Journal of Political Economy* 70: 1–13.

I'm mangling the title. But his point is that, if you can't spend more money than you have, then, as prices rise, people are going to buy less. And it almost doesn't matter how individual agents make decisions. The fact that you can't spend more money than you've got means that as the price of some good increases, the amount of what you can buy is going to decline. Demand curves will slope the right way.

You can make a similar argument about supply curves if you can't stay in business while losing money perpetually. And Gary Becker turned that into: this means that we can just have neoclassical optimizing agents all the time because it doesn't matter. But what he was really doing in that paper was saying, they're just these institutional constraints about exchange. A lot of what we think of as supply and demand is going to come from those institutional constraints on how exchange happens; the individual decision-making method mechanisms don't matter and you can abstract away from them. That's the real message of that paper.

We're going to have to do a lot more of that to—say, under these institutional constraints, whole very different ranges of individual decision-making will give us this kind of macroscopic phenomenon—try and pin down what are the limits of that or what are the ranges of individual decision-making mechanisms, which in a certain institutional constraint will give us these macrophenomena.

That would be a productive way forward. Thinking about spin foams and can we somehow map differential connections onto interest rates? Probably not so much.

M. DEL RIO-CHANONA I have to say, I agree with you, Cosma. Maybe you probably don't want me to agree too much.

[COSMA LAUGHS]

I'd say we have to stop copy-pasting things from physics into economics. We have to stop reinventing the wheel. We have to start reading the economics literature, because if we want economists to listen to us, well, we might as well read their literature, right?

C. SHALIZI Totally fair.

M. DEL RIO-CHANONA Yes. Now, since you want me to disagree a bit, I'll go with only a bit. First, let me rephrase your question. You say there are many advances in physics. Can we use any of those in economics? Well, let me just say there are many new things right now. The world is evolving. One of the great revolutions is computer science, machine learning, the amount of data availability. So, I'd say, let's focus on that, and let's see what we can use to do economics better.

Now, that's when I thought about Ole Peters. You gave a great talk yesterday. I loved it. I particularly loved the experiments.

But one of the things that surprised me was that you said, well, now we have this new mathematical framework. Let's go back 300 years ago and do it better. Can we really think, just with math, because the math is sort of something that's always been in our mind—can we go back 300 years later and do better than all the economists have?

Your results suggest yes. But I'd say maybe that's harder than focusing on today, seeing what are the new things right now—what's new from physics, what's new from computer science, what's new in data—and doing better for the future. That would be my question. What makes us different if we're just putting some more math and going back?

O. PETERS Right. Well, the good thing is that we are several people and we can all do different things. And we really should do that. People should go and use machine learning and build large agent-based models and explore what needs to be explored.

And I think many of the most pressing policy issues, things where you want to build an artificial environment to try out some new protocol, basically—we need those systems, but not everyone has to work on them.

So, I sort of take the freedom to say: I cheer you all on and I'm grateful that you're doing that. I'm grateful for what everyone is doing in their research. We're all trying to do our best. But I've picked this one approach—or the approach has picked me—and it's been surprisingly successful.

So, when you ask why in the world it would help to go back 300 years, and what makes us think that we could do better, I think the answer is just path dependence. You laid foundations 300 or 350 years ago with the tools that were available at the time. Our tools have evolved dramatically, so it does make a lot of sense to use the best practice knowledge that we have right now, and just try to go back and see if any of those were more or less answered, or if it depends. Some people consider these questions answered with the tools of the time. Other people have sort of a tummy ache when they look at the solutions, and for a few of them, there's something conceptually circular, and so on.

So, I think that's the answer, right? Mathematics has evolved dramatically. I agree we should read the literature, but even there it's not a yes–no answer. Sometimes it's good not to know much about a theory, because that makes it more likely that you'll develop a radically new perspective.

I looked up the citations on Kahneman's paper from '79, and it's something like 350,000 citations. So, look, you can't read that literature. It won't work. But it can be one of those gatekeepers; you make a contribution to a problem, and you're told, "Oh, you haven't read the literature." And you say, "Well, of course I haven't read the literature, because the literature has millions of papers. It's not possible to read. This raises questions: How do we communicate in science? What are our communication systems? How do we filter information? That general problem is maybe particularly bad in economics, but not *only* bad there.

D. WOLPERT I agree with Cosma. I've worn many hats and I have actually been published in game theory journals several times, as well as physics journals. I would say that there's an interesting cultural issue here. Economics has been a bit of an easy target to kick in this entire meeting. I would say physicists are arrogant and economists are smug.

[LAUGHTER]

I can say that because I've been both. And what I mean by that is that physicists think that their models can solve everything. That was actually the major, major problem with the SFI in the early days. It was, "Get out of the way, I don't care what field you are. We are here now, time to revolutionize, end of science, etc., etc."

And a pox upon the house of econophysics. It's starting to come around, but it still doesn't come close to fully appreciating what's been done in conventional economics. There frankly is a hell of a lot of value in the game theory and economics literature. There's a lot broken about it, but there are also lots of actually brilliant insights. Gary Becker was mentioned—he certainly did a lot of fascinating work in this literature.

I agree wholeheartedly that you don't want to know about a field when you are starting to work in it for the first time. You want to be able to be fresh. But once you've gone through that, and you've got your sort of general stance set, then you've got to learn, well, just what did the giants whose shoulders I want to stand on have to say before me, even if I think what they said is somewhat wrong.

Sometimes it's good not to know much about a theory, because that makes it more likely that you'll develop a radically new perspective.

Now, those are the difficulties with physicists moving into economics. There's another problem over on the economists' side of the fence, a huge one. There was actually a very sobering study—I wish I remembered the details—about five years ago, where somebody went to petitioners in many different domains and said, if there were a new technique that was not from your field, but was making a lot of buzz in other fields, how likely would you be to learn about it and to try to integrate it into what you're doing? Economists were the worst. They were the bottom.

The big advantage of physicists—I think Doyne Farmer may have once said this to me—is not what they have learned, the tools. It's *how* they have learned to think. In particular, physicists are quite good at being very, very broad, taking tools from all over the place. That is something that economists are very, very remiss in. It is a b—tch to try to get published in game theory. I have gotten many rejection letters that say things like, "Oh, well, your arguments are based upon Bayesian statistics and I don't believe in Bayes's theorem."

[LAUGHTER]

What do you do? Well, you're sort of stuck. So I feel your pain.

O. PETERS My thing is over. I'm not doing it any more.

D. WOLPERT The sailing of the seven seas, you have now entered into port—all kinds of metaphors here. That problem is actually a broken cultural aspect of economics as a result. For example, there was mention of Tom Sandholm's work earlier. I think Rob may have mentioned that. Some of the best work is actually occurring in what's called the UAI community: Uncertainty in Artificial Intelligence. Some of the best work on game theory is actually occurring there and so on. So, part of this is actually the different cultural stances. To make all of that sort of vague blather somewhat more concrete, I would actually follow up with some of the things that Jessica was saying earlier today.

It is a cliché in economics that the farm, the economy, what have you, is the computer. Sure, mother's milk. We all believe that. So far as I know, and I have looked at this literature, there have been essentially zero papers out there that actually tried to make that mean anything other than triviality.

There are lots of words. You'll find tons of words. There are lots of teeny little systems where you've got several agents that have very, very restricted move spaces, and here's something about how the Nash equilibrium shifts under this, that, and the other condition that you can actually solve. That's the team game literature; I'm thinking of particularly going back to Radner and so on.

But there's been nothing where people have tried to say, "Let's look at a large social organization," and view it not as an information transmission system, because that's all that Shannon's information theory can give us.

We don't care in the economy about just getting information from point A to point B. We care about transforming it, performing a computation on it. That's what the magic is. That's part of what Jessica was going after. Information theory in the Shannon sense—just transmitting information from A to B—there is some of that in game theory, in economics, arguably not as much as there should be, but that's certainly nothing new.

But the whole notion: What would it even mean to view a social organization as an information-transformation system? What bodies of mathematics should we call upon to try to actually start to grapple with that? That is where I think the cultural stance of physicists can actually have a very powerful role to play.

I don't think the tools can. Using spins—God love 'em, but they're not going to answer that question. But this sort of omnivorous eclectic mindset, which is very much what the SFI is, that is an aspect of physics that has been built into the genome of the SFI. That's what would be necessary to be able to go after those kinds of aspects of conventional economics, which are just sitting there and have been for over half a century. Nobody's actually tried to do anything with them.

D. C. KRAKAUER By physics, I mean approximation, compression, mechanism, optimization. Data science is not that. Data science is not science. I think it's important that we make that statement. SFI doesn't do data science. SFI does science.

[LAUGHTER]

I do want to ask a second question that David raised, which has to do with institutions and what it means to think economically in a world where we have little greedy Turing machines running around establishing contracts on our behalf.

I think that's a really interesting idea: the co-evolution of economic thought and technology. Before, we had markets, but we didn't have economic thinking in the form we do. There is a lot of emphasis in this meeting on institutions. My notes are full of the word "institution," so clearly SFI has to start thinking very carefully about them. Would folks be willing to make a remark about the co-evolution of economic thought and the kind of constraints that you've all referred to in what we would call institutions?

M. DEL RIO-CHANONA Sorry, could you rephrase the last part?

D. C. KRAKAUER To what extent is the theory that we're developing contingent on the structure of the economy? The economy is a technology of a certain kind that enables transactions, and it's not present in the theory, as I see it. You write down a simple bimatrix game. You write down your time-average behavior. Where is the actual substrate of the economy on which that dynamic plays out, and how will that influence the evolution of the theory?

D. WOLPERT I can say a few things. First, actually, just a warning: the word "institution" means two things in the economics literature. One of them is actually much closer to what we would mean by social norms. The other one is much closer to what I think what we would all mean by the word "institution."

Institutions, economic systems in general, going all the way back, even arguably to barter money—those are technologies. People refer to financial technologies, and it's more than just financial. It's also more generally economic technologies. The invention of writing was, in a certain sense, an economic technology and an invention that greatly facilitated cooperation, not just—to make a connection with what Eric said—in the conventional two percent of the visible matter that we see, but also obviously writing had a great impact on the ninety-eight percent of the dark matter of hidden cooperation as well. So there have been these technological advances. There have been extraordinarily few, if you think about it. We still run firms by basically doing case studies and narratives and just thinking about, "Well, I don't know. How do you think that the most recent brilliant CEO did things? Maybe we should emulate them."

. . . we are actually living in the aftermath of the singularity. The singularity is what we call *the Industrial Revolution.*

The technology we're using to organize ourselves goes back many, many thousands of years and hasn't changed much. Institutions are perhaps the only overlay on top of the underlying human inherent nature that is modifying that at all. But I don't think anybody's approached it as a technology, in the way that you're describing, to try to see how that technology's innovations have developed, and how it should develop into the future.

I mean, obviously the first thing that I talked about was all about new institutions that might be forming before we're ready for them.

C. SHALIZI There are people who've looked at that, though, David. They're historians of technology and organization and institutions.

D. WOLPERT True. But nothing quantitative.

C. SHALIZI Some of them are pretty quantitative. You have these books like *The Visible Hand*. You have books like *The Control Revolution*. You have JoAnne Yates's studies of how, in the nineteenth century, you developed all of these organizational and paper technologies for coordinating very large numbers of people, for literally making the trains run on time, which were then adopted by corporations and other parts of the economy, and in part what made it possible to have a corporate economy.

You have people who have studied evolution of institutions, both the informal and the formal variety, from an economic evolutionary game-theory point of view.

You have Douglass North, who died just a few years ago. You have all of the people around him. You have Peyton Young, who's been a frequent visitor at the Institute, who has this lovely book about how institutions can evolve from game-theoretic interactions.

Very few of them, I will agree, have thought about how you would do institutional engineering to come up with, for example, if you want to get institutions of a certain form with certain outcomes, what would you need to do that connects to some of the mechanism-design stuff that I guess Rob was talking about in the previous session.

But there are people who have been thinking about this.

D. WOLPERT We should talk.

C. SHALIZI Yeah. And the fact that—most of them would not put it this way, but we are actually living in the aftermath of the singularity. The singularity is what we call *the Industrial Revolution*. The artificial intelligences which have taken over the world are called corporations, governments, and the global market.

And they are instantiated as a distributed processing system on human beings. They don't particularly care about human beings. Their motives are inscrutable, to the extent that they have any, except that they'd like to get bigger. They've run on people. They're implemented on people. But they are not people. They are not persons in any meaningful sense, but they are computational processes. They have their own feedback loops. They try to maintain themselves. They tend to expand. And I'm sorry, we're living in the post-singularity, post-apocalyptic nightmare.

[LAUGHTER]

M. DEL RIO-CHANONA There's also the evolutionary economics literature. You have Metcalf, Montolio, Dulce's model—those are a great push. I wanted to make just two comments. One is, for technological advances, as everything, you always need a variety of things. So I'm really happy to see people working on different types of things.

The other thing is—who mentioned data science? I said, we need to go into data, and I'm citing Doyne Farmer when I say we need to gather all the data. Because unless we observe the real world . . . and, as you say, as physicists, I think we need to abstract the world. But I don't see how we're going to observe it without data. I agree—data science is not a science; it's more like pretty plots. But yeah, I stand strong when I say we cannot hope to understand the economy if we don't understand the data.

ERIC D. BEINHOCKER David, you said you could characterize the economy as a technology to facilitate exchange. I would say that's too narrow a view. A broader view would say the economy is a set of technologies to facilitate order creation in physical and social order.

Now, this is an old perspective—Hayek had it, Simon had it, Georgescu–Roegen had it, but it's always been just words and never really been carried through. Physicists know a lot about order-creation, both from an energy/thermodynamic perspective and from an informational perspective.

Veblen back in the 1920s wrote a famous article, "Why Isn't Economics an Evolutionary Science?" I would ask, why is it also not a thermodynamic, information-theoretic science? You know, essentially a system of free-energy minimization within constraints. Discuss.

[LAUGHTER]

O. PETERS I would just like to make a comment on the economics evolution or economics ecology analogy. I have learned this from [SFI Vice President for Science] Jennifer Dunne back there and I was surprised that you didn't mention it, but I love this circularity. I think both Linnaeus and Darwin used the term "economy of nature" when they talked about ecology.

So, I don't know what this means. Either this is just complete hopelessness—to understand the economy, we look at ecology, and to understand ecology, we look at the economy. It is completely circular. Nothing happens. And of course the optimistic vision of that is we have two fields that have different methods and they speak to each other. But I just wanted to mention that there's this amazing ping pong between these two concepts that sort of goes back every couple of decades.

AUDIENCE MEMBER Going back to the data, I have a question. Yes, there is a wealth of data out there, but the data quality in physical systems versus social systems is quite different. I actually had the privilege to work with data in heliophysics and radio astronomy and also in biology, but also working with data in social media, economic data, and so on.

The data quality is really different. In physical systems, it's measured by instruments. It's very clean, beautiful, nice to work with. In social systems, it's very noisy. Eighty percent of the time, what you do is just clean up the data. Right? So, going back to this idea that data informs empirically different methods and methodologies in physical systems versus in social systems.

To give you a direct example, the FAA, for airline-safety data, uses both instrument data, measured on the ground and on the planes, and data collected from the pilots—the reports of the pilots.

The models of airline safety—which are really important, right?—from the pilots are very different from the ones measured by instruments. So, the fact that we actually don't crash is because these models are composite, between the physical systems and the reports of the pilots. So, I was wondering what would be your stance with respect to the empirical side of social sciences and physical sciences as we are thinking of them as co-evolving in this space?

M. DEL RIO-CHANONA I agree those differences are between the social science data and physical data. I think data is improving. I also agree with Cosma when he says we cannot just copy–paste; we cannot take things from physics and apply to social science. We have to think of new methods to analyze data that's a lot noisier.

I am optimistic that it's improving. The thing is, you said eighty percent of the time it's data cleaning. And I think people tend to look down on gathering data, collecting data, cleaning the data. That's not science. Science is writing equations. Well, I'm sorry, but when you're a physicist, part of being a good physicist is designing the experiment. Doing it correctly. Right now, part of being an economist is going to be thinking of ways of gathering the data.

And I think it's scary. We shouldn't leave the blackboard, of course, but we need to start thinking that there's also a lot of pride in collecting data of good quality. If you find a way to go from noisy data or to clean it up or to find a mechanism to actually understand it, that's a major advancement.

So, we shouldn't look down on the data problem. I think that's a fundamental part of it. It doesn't replace a fundamental part of understanding the system of doing models and actually doing science, doing complexity. But that's the point.

. . . we need to start thinking that there's also a lot of pride in collecting data of good quality. If you find a way to go from noisy data or to clean it up or to find a mechanism to actually understand it, that's a major advancement.

C. SHALIZI I will agree with Maria that the data-gathering part and the data preparation are essential. There are people who spend their lives gathering and compiling and cleaning social data, and especially economic data. They work for official statistical agencies and they are the people who figure out things like unemployment rates, labor-force participation, investment rates, GDP, all of these official statistics that, at least in developed countries, we've come to rely on and find fairly useful and reliable.

A lot of the new data becoming available outside of what we get from the traditional official statistics, is stuff like Twitter or people's usage of particular companies. As you say, it is incredibly noisy. If you try to actually wrestle with it, it is a pain, and eighty percent of the time on data cleaning is very optimistic.

REMEMBERING THREE SFI THINKERS

Ole Peters

A discussion of physics and economics would be incomplete without devoting some time to the relationship between theory and data. I will do this by remembering three thinkers I met at SFI, all of whom passed away recently: Murray Gell-Mann, Ken Arrow, and Reuben Hersh.

Murray's mind had a quality of pristine disinterestedness. I once pointed out to him a flaw in an argument, which invalidated a statement. Murray beamed with delight: like noticing a dripping of brake fluid, spotting such flaws is essential to preventing future car crashes. But he raced on: that long-held beliefs can be wrong was one of his long-held beliefs. But why? What assumptions would one have to make for the belief to be right, and could those assumptions be justified, did they sometimes make sense, could we draw a line around the space of all assumptions that make the false statement true? Playfulness without losing sight of the core problem: yes, the statement was false. But still...

I remember a disagreement with Ken. Ken had chosen a particular regularization of an expression which was mathematically ill-defined and, in a sense, up for grabs. I failed, despite Murray's help, to convince him that another interpretation of the expression was not just valid but would be more fruitful, as it is in physics. We made no progress for two years, until I suggested we set aside our formal disagreement and explore the consequences of this different interpretation. We looked at the data—namely, we looked at the model world created by the alternative interpretation—and saw that it contained structures we knew to exist in real economic life. The data unlocked something where our formal back-and-forth had got stuck.

Computer-aided proofs shone an early light on what we're grappling with today: is a proof a proof if no human can complete it?

Finally, Reuben. Like Murray, he had a fluency in switching perspectives and turning a concept on its side to see another angle. Acutely aware of the social dimension, he warned against "Mathematics as rhetoric." His thinking connected the very abstract with the very real, personal, subjective. Mathematics not as something to be discovered, nor invented (echoes of the quark), but as something collectively held: a social reality. Computer-aided proofs shone an early light on what we're grappling with today: is a proof a proof if no human can complete it? The production of publishable papers with the help of computers is part of our social reality. How much of it is science? Reuben's focus on meaning was like a vaccine against mistaking a collection of summary statistics, say, for insight into the workings of a system.

One of the potential advantages of it is that it is very fast. It appears at a much higher cycle time than the official statistics. So, the problem we would have would be how do we use a very large number of very noisy measurements of whatever it is that we're interested in to inform our models.

That is a statistical problem. Or as we now say, a data-science problem. It is something where there are techniques for how to use multiple imperfect proxies to help measure a latent construct. That's a methodological problem where people can import ideas. And there'll probably be some new ones based on the exact types of data that we're dealing with.

But institutional, professional rewards for data collection and improving the quality of the data available—yes. That certainly has to be part of our professional systems. Otherwise we are going to gather horribly biased, noisy data, pretend that it's real and entirely accurate and do all kinds of awful things with complete confidence until it blows up in our face.

JOSHUA EPSTEIN Yeah, I had a comment on this whole issue of data and science. There are counterexamples, but, by and large, science does not progress from data. For the big advances in physics—the theory precedes the data. Maxwell's equations said, holy mackerel, there are going to be radio waves, and Hertz goes out and makes that measurement. Einstein and relativity—light should deflect in a gravitational field. Eddington makes that experiment. The theory tells you what data to collect, and I think that's really important. I mean, the data don't tell you the theory. In fact, all the great idealizations of physics would never be observed in any data. You'll never find data that give Earth and planets as point masses or perfectly frictionless surfaces or actually ideal gases or anything else.

The game is to idealize and then use data to see if the idealization is fertile. On Eric's point, Samuelson was very explicit. If you read his Nobel Prize lecture, he's very explicit about basing the entire thing on thermodynamics and it's not clear that that's the most fruitful idealization. I actually think it's not, but it's certain that it was the same impulse.

So, I think this whole business about data first is really kind of misguided. I agree with Ole on that. But I disagree with you that Newton's laws aren't extremal; $F = ma$ does minimize action after all.

. . . by and large, science does not progress from data. The big advances in physics—the theory precedes the data.

O. PETERS Of course. Yes. My conceptualizations of it came later, the maximization. I would like to push back against this caricature of data being easy and perfect in the physical sciences and very, very noisy in the social sciences. My background is in statistical mechanics, and the Ising modeling has come up.

It's something I've never used to describe anything economic—who knows if that's useful ever. But that's a model of phase transitions, and in phase transitions it's one of these cases of having something that is extremely difficult to measure.

The science history around the 1930s and 1940s is extremely interesting if you want to know about how difficult it is to actually verify or falsify some theory in terms of data. There was a theory developed for phase transitions. The mean-field theory was there in the early 1930s, and the experiments contradicted the theory.

The theory was so simple and beautiful that everyone believed the theory, and the experiments were not strong enough. Looking back now, the data say the theory is wrong, but back then that was not enough to recognize that it was wrong. It took the analytical solution of the Ising model in 1944 to show that mean-field theory was not exact, and that in fact the experiments had been better. So this ping pong back and forth between theory going first and the data coming in and correcting it—this is everywhere.

J. EPSTEIN What about Karl Popper? His vision of science was called *Conjectures and Refutations.*

M. DEL RIO-CHANONA All right, so, I'll just quickly answer. Yes, theory, ninety percent of the time, has to come first, though data can tell you which theory is right.

J. EPSTEIN But that doesn't tell you a theory. It adjudicates among competing theories.

M. DEL RIO-CHANONA So, just an anecdote—there was an argument between Poisson and Fresnel. The thing that proved it right was the experiment. The other thing is, when we go to theories, we usually derive them either from a mathematical perspective which we have been trying for 300 years, or we do it from observation.

Believe it or not, there's a story about Newton. The apple fell and he thought, "Okay, this should be a law." So, I'm going to ask you, how do you observe the economy? We've been trying to observe it for 300 years. And I believe the way to do science is, you observe the real world, and then you try to develop the theory.

My point is, I think we're reaching a limit of how much we can observe with our eyes and how much we can observe with our daily interactions, and I think the way of doing observations is with data. I might be wrong.

DARIO VILLANI A lot of spending time, like they were saying, the ninety percent of time cleaning data is also because a lot of the frameworks that people use are extremely rigid. The reality is that, if you use techniques that are much more statistical, then you are much less vulnerable to bad data feed and techniques that are nowadays really adaptive, like humans. If my son is going to cut his hair, I'm going to still be able to recognize him. So, the distortion caused by bad data feed and how scared people are of that, it's really related to top-down modeling and measures that are extremely hard to know. But there is a way to be less vulnerable to that.

C. SHALIZI I will push back a bit. As a statistician who does this all the time, even if you're using the best methods we've got, the raw form in which you are getting the data from something like survey research or Twitter or purchasing records contains a lot of stuff which is just nonsense. And it has to be cleaned or else it's either not even possible to deal with it computationally or you are just fitting a lot of the errors in the recording process.

D. VILLANI I like that you assumed I'm not a statistician, it's funny.

C. SHALIZI Ah . . . Sir, I don't know you.

D. C. KRAKAUER Are you insulting each other by calling each other statisticians or complimenting each other?

[LAUGHTER]

C. SHALIZI We are bonding in the way of our people.

D. C. KRAKAUER You're bonding. But anyway, thank you very much to the whole panel.

TATTOO

THE ECONOMIC ORGANISM

[Panel Discussion]

Moderated by William Tracy,
featuring C. Mónica Capra, Scott Page,
Rajiv Sethi & Geoffrey West

WILLIAM TRACY David got us started off thinking about areas of complexity economics that have been inspired by computation, areas of complexity economics inspired by physics, and then a much broader pool of areas that have been inspired by biology. Across this biological spectrum there's a lot more heterogeneity in what people are actually doing within it, so much so that we had many more people. We ended up splitting this into two different panels.

For the panelists in particular, I'll encourage you to follow David Wolpert's very good example from the last panel of telling us what you do and then closing, not by trying to defend any aspect of biology, but by sharing what it is that you think is the most important area for people interested, not necessarily in the discipline of economics, but in the analysis of complex economic systems. What area is most important for us to develop next? And with that I'll turn it over to Mónica to kick us off.

C. MÓNICA CAPRA Thank you so much. One issue is how to narrow the gap between what the scientists here, for example, at the Santa Fe Institute, have advanced in terms of understanding complexity and especially complexity with regards to the economy, and closing the gap between that and what the general public or the society of economists can do.

OPPOSITE *Economic Curiosity: The Man Who Sold His Face*

I think we need a set of measures and a language that can give us some type of label to the measures. Can we come up with a common understanding of what it is that we mean by the level of complexity in the economy and the dynamics of adaptation? Having those measures linked to specific labels is important. Let me give you an example.

There's work by Hausmann and Hidalgo on measures of economic complexity that looks at the development or the growth or the economic value of an economy based on how many links it has, with respect to trade, with respect to how the firms are linked with each other, and so on and so forth. In this account, we have a specific measure of complexity and we can do something about it. So, moving from measurement to labeling so that we can have a common understanding of what it is that we are talking about—super important. Again, going back to the presentation yesterday to create a common knowledge about a new kind of mindset of seeing the economy.

Can we come up with a common understanding of what it is that we mean by the level of complexity in the economy and the dynamics of adaptation?

Brian talked a bit yesterday about seeing the economy as an evolving system and not a static, equilibrium type. That's hard to do if we don't have that kind of measure and label. The other thing is that our brains have evolved to deal with simple kinds of linear environments. It's going to be a challenge to help consumers adjust to the complexity. When we talk about narrowing the gap between what the scientists have been able to develop in the last thirty years and what the decision-makers can do, we could think of the decision-makers as the consumers, organizations and firms and the government. Again, I think that consumers in general face a very difficult time figuring out how to navigate this complexity.

Perhaps what will help is some type of augmented intelligence, right? That will help people identify misinformation, for example, or try to figure out how to make better decisions in a complex world. For governments, I think that what will really help is to have some type of tool in which you can visualize quickly how different policies can impact the ecosystem. I'm just thinking aloud. For example, if you have a local government interested in incentivizing entrepreneurship, an ecosystem of entrepreneurial activity, then you could use these kind of simulated worlds to see, well, what happens if we change the tax structure? And then you can have a visualization of what will happen. In addition, a visualization can not only generate these possibilities and counterfactuals, but also it can help in evaluating the effectiveness of the policy so that policymakers can have quick feedback in terms of how their policies are doing.

With respect to firms, I think that it would be really useful to think about how to use technology to do experiments. Matt Jackson gave a very nice example of a field experiment where you can actually kind of decide how to do an intervention based on the knowledge you have. Today, we have an amazing ability to collect very fine data. So, organizations could do experiments like . . . I'm just thinking of an example. Suppose that you have Uber or Lyft and there are different ways in which they can incentivize people to keep their workers. It shouldn't be that difficult to kind of come up with an experiment and collect data. Then, based on what you collect, you can inform businesses to make better decisions.

SCOTT PAGE Great. I'm Scott Page from the University of Michigan. The question here is how we should think about the economy. To paraphrase Brian a little bit, I guess I could say John Holland is the answer. What's the problem? I want to take a very pragmatic approach to this. If you're in a business network, whether you're in government, for-profit, nonprofit, whatever, the reason you care about a complex-systems perspective is to be better at what you do. Like, how do I make predictions? How do I guide actions? How do I design things? How do I construct teams, organizations, institutions in order to do those things? To me, the question on the table about Santa Fe has always been, does this complex-systems approach—the stuff Josh does, the stuff we saw Matt do, the stuff that we saw Jessica do, these things that we get up every day and try to do—does that help us at a pragmatic level? I'm not a biologist; I don't even play one on TV in the classroom.

I want to go back to some stuff Jessica talked about coarse graining. You could think about the equilibrium models that we use as strategic, rational, coarse graining of reality, and complex systems models have a different coarse graining. There's a wonderful book by Anne-Marie Slaughter called *The Chessboard and the Web* that plays this out very nicely. You can think of international relations as a chessboard of strategic actors. You can also think of it as a global web of players, including multinational corporations, nonprofits, that sort of stuff.

The point is not that one of these is wrong and one of these is right. It's not that my model beats your model. It's more a both/and thing. Some of the things that markets have done are fantastic. I mean, look at these socks, for crying out loud.

[LAUGHTER]

But it's also the case that there are some blind spots. I thought Eric did a wonderful job of raising the fact that if we stick to one model, we're bound to fail. There's the famous George Box quote that all models are wrong. Then the question becomes, how do we think about developing tools, procedures, things for people who work? If I go to the congressional budget office or if I go to Exxon or if I go to NASA, where I'll go next week, how do we think about taking this science and making it a practical value for them? What tools do I give them? What measures, what techniques? Bill Rand, I wish I could clone—Bill was a student of John Holland's and mine. I wish I had 900 Bill Rands I could just send out there to make the world a better place.

But that then leads to the question of where is the real value add? It's not a bolt-on, but there's a dialogue between this chessboard and this web framework. What worries me at the moment is, because there is so much data, there are certain things we're doing that are narrowing how we look at problems. We're not being very expansive. Let me give just one sort of example. There's a lot of AB testing that goes on, right? Like, let's try this and see whether it worked. A lot of the nudge research is just simple testing. But that ignores what Jessica was talking about, and Jennifer Dunne speaks about these sort of ecosystem constructions, these web constructions.

The reality is that it can often take a long time for a system to learn, given a new set of incentives, structures, information. So, the simple example I give all the time is, in 1981, the Los Angeles Lakers were one of the best teams in the history of the NBA. They won sixty-seven games. Magic, Kareem, James Worthy. The three-point shot had been introduced two years before that in 1979. For the year, they made thirteen. For the year. Now, it's not complicated math—Michael Mauboussin and I figured out that three is fifty percent more than two, right?

Last year the Houston Rockets made 1300. This is thirty years later. Why? Well, they had to change their offenses. They had to change how they learn. They had to learn how to shoot that shot. Right? When you think about doing an AB test, one of the things about having an equilibrium, rational view is you assume instantaneously you sort of climb that gradient and go to where you're going to go. As opposed to, as Josh would say, Agent_Zero may take a while, right? The messiness of Agent_Zero might take a while to sort out.

Last thing is the standard approach, and this—I'm going to channel Beinhocker to a large extent—is siloed. One of the things we're doing in Michigan is this thing we call "the choice." When you think about making an institutional choice, you can choose between markets, hierarchies, democracies, sort of self-organized things, and now algorithms. It's really almost like having an institutional DNA. We use markets here, hierarchies here, democracies here. Any one of your organizations is probably having markets in places, voting on things in places, hierarchical chains of commands in others. Those things create huge spillovers in our knowledge, in our heuristics, in our behaviors, in our levels of trust, in the information we have.

We need to think about using these sorts of models. What are the spillovers from these institutions? Then Josh can tell us what we need to measure. And then, once we measure them, the map. I thought it was beautiful, the network stuff showing us that how these different institutional choices make completely different network structures. And then we can see how the institutions affect what society is and how what society is affects how these institutions are going to work. And I'll pass it to Rajiv, who's going to explain how to do it. Seriously!

RAJIV SETHI Thanks, Scott. Since we are all paraphrasing Brian Arthur, let me start there also. If I understood correctly, one of the main points that he made yesterday was that the essence of complexity economics is nonequilibrium reasoning, or, if I can put it slightly differently, disequilibrium dynamics. One way to do this is through the agent-based computational approach, which was discussed in the first panel this morning. Now, if one takes that view of complexity economics, then, unfortunately, over the last thirty years, the impact of complexity science on economics has been really quite limited. It hasn't made its way into the graduate curriculum. It's not published in the leading journals and it's still peripheral, I would say, to the discipline.

You have some of the best practitioners right here—Rob Axtell, Josh Epstein, Blake LeBaron. I would add Leigh Tesfatsion at Iowa State to that list. There are people who are doing extraordinary work with this, but it's really not on the radar screen of graduate students or most professional researchers either. So let me just try to begin by asking why. Why is this?

The cynical approach, I think, would be the two-mountain view of David Colander that Brian Arthur mentioned yesterday, which was that the profession has climbed the wrong mountain and it's too much trouble to climb back down and start again. So we are stuck where we are. This actually this reminds me of Stuart Kauffman's piece in the original volume on optimization on rugged landscapes; you know, you just end up where you are and it's too much trouble to change.

But let me give a less cynical view. I'm talking here from the perspective of editors and referees of the standard journals. My view of agent-based models is that they're immensely promising, but they're too easy to construct and too hard to evaluate. The insiders like Josh and Rob and Blake and Leigh can tell a robust model with deep insight from one that is fragile and very sensitive to changes in specifications. But the outsiders can't. And, until this thing takes hold, the referees and editors are outsiders. They need to evaluate this and they just can't do it. It's not that they dismiss it out of hand as worthless; it's that they don't know whether it is worthless or not. The only way that that will change is if we have an exemplar. When we look at information economics, we think of Arrow and Akerlof, Spence and Stiglitz, and these

ENTROPY & THE ECONOMY

C. Mónica Capra

From the point of view of SFI scientists, mainstream economists' reluctance to embrace complexity in their teaching and practice must look absurd. After all, economies are complex adaptive systems and should be analyzed with the appropriate tools.

> If one were to put on a physicist's glasses, what role would entropy play in our understanding of complex economies?

This and other panels argued that mainstream economists, who disproportionately influence policymakers, seem stuck with outdated equilibrium-centric tools that date back to Alfred Marshall's early twentieth-century work. The panel discussed different explanations for this conundrum. I argued that the culture in economics does not reward innovation. To be precise, until recently, most economics departments did not have post-doctoral programs. In addition, faculty are often discouraged from collaborating across disciplines and co-authoring papers. These practices have prevented the bottom-up and transdisciplinary flow of knowledge that is needed to keep pace with scientific advancement. Yet the panel found value in the traditional equilibrium-centric approach, despite much warranted criticism. Traditional economics emphasizes parsimony and is meant to create a simple abstraction of reality. Economic models that are abstractions are not meant to be tools for making precise predictions of what will happen in our complex world. Instead, these models are frameworks for thinking.

In relation to this latter issue, Geoffrey West's comments on noticeable deficiencies in economic thinking intrigued me. In the panel, West mentioned a puzzling absence of basic physical properties such as energy and entropy in economics textbooks. This absence suggests that economists' thinking is divorced from reality. Indeed, nothing exists without energy. Following West's example, I searched for the keyword "entropy" in prestigious economic journals. I found several studies that incorporate ideas of entropy into the economics of information. However, it seems to me that the economists' approach is only scratching the surface of what West may have in mind.

If one were to put on a physicist's glasses, what role would entropy play in our understanding of complex economies? Geoff has successfully introduced ideas of physics, such as scaling laws, to characterize cities and understand their evolution. I hope he scales up his framework to include intercity interactions such as trade and commerce. I also agree with Geoff that economists have yet to wrestle with time. How can we integrate the acceleration of time into our models and measurements? There is much progress to be made and questions to be answered, but I am left with some degree of certainty that Geoff's framework of thinking is where we need to start.

Further Reading

West, Geoffrey B. *Scale: the Universal Laws of Life, Growth, and Death in Organisms, Cities, and Companies.* New York: Penguin Books, 2018.

are exemplars. This is the way it's done. People know what's good and bad by taking this as a benchmark.

When we look at causal inference in empirical work in economics—you know, Card and Krueger and Josh Angrist's work—we say, "Okay, this is it. This is the gold standard. This is how it's done." You can then use that to benchmark submissions. We need an exemplar. It will happen, and once it happens the floodgates will open. I really do believe that this methodology is very fertile, but unless we have a way of setting and establishing standards, it's not going to have much impact in the graduate curriculum and in the research profession.

I have two minutes left. We were asked to review tools, so I can talk about this in the discussion if you'd like. Let me talk about a tool that I have been developing as part of a team.

. . . the Santa Fe Institute should be involved in the ultimate pragmatic problem: the future of this planet and saving this planet. . . . I think one of the great disgraces of academic economists and economic departments is that they're not at the forefront of this, already leading it.

It's an IARPA-funded project.[1] It's a large project headed out of the Information Sciences Institute at the University of Southern California. It's headed by a computer scientist called Aram Galstyan and it's mostly computer scientists. I think I'm the only economist; there's a psychologist and a few other folks. But the basic idea is it's the IARPA follow-up to ACE, the aggregate contingent forecasting competition that was won by Phil Tetlock and Barbara Mellers's group, The Good Judgment Project. This was a competition to see whether or not teams of forecasters could out-compete the intelligence community analysts

1 The Intelligence Advanced Research Projects Activity (IARPA) is an agency within the US Office of the Director of National Intelligence.

who have access to classified information. What The Good Judgment Project did was remarkable, right? They out-competed them year after year—thirty percent better forecasting rates and, you know, the rest is history.

Now what IARPA wanted to do is to say, "Look, sometimes you have to make quick decisions. You can't just have humans make them; you have to combine human and machine methods." And so the new competition of which I'm a part is called HFC: the Hybrid Forecasting Competition. It's really trying to see if we can combine human judgment with machine models to generate rapid forecast for large numbers of IFPs—Individual Forecasting Problems—and see if we can beat the purely human forecasters. It's still at an early stage. But the reason I was brought into it—I'm a small part of this group, but what I've been trying to do is to build what we call a *hybrid-prediction market*.

The idea is to take a machine model forecast and to have a trading bot that basically acts as if it believes that forecast, has a budget, has risk preferences, and places orders to buy and sell based on those beliefs. Those orders look a bit like the algorithmic market makers that you see on regular markets, but they're not market makers. They're trading on a fundamental strategy; they really believe that forecast. Then to have a bunch of humans coming in and trading with these bots and with each other, and trying to see then whether the market-filtered forecast can out-compete the pure machine forecast. We're at a very early stage of this process. I'm out of time, so I'll stop there.

GEOFFREY WEST My name is Geoffrey West and I'm a physicist. All the things I thought about saying a couple of days ago have been said, but I want to emphasize a few things.

Taking off from what Scott said about pragmatic problems and the role of the Santa Fe Institute in complexity science, I think that the Santa Fe Institute should be involved in the ultimate pragmatic problem: the future of this planet and saving this planet. That is the ultimate pragmatic problem, and I think one of the great disgraces of academic economists and economic departments is that they're not at the forefront of this, already leading it. I find that astonishing, frankly, and I take as my point of departure the fact that it was physicists at the beginning of the Second World War who led the battle to try to develop a terrible weapon that would somehow give us the ability to stop the Second World War because of the threat of authoritarian rule and fascist governments. You know, the threat we face now, I think, is many orders of magnitude greater than it was in 1941. And a large part of that threat could be thought of as under the guise of economics in its greater sense. I think it's morally reprehensible that the economics departments of this country are not leading the way. All those fancy economists, by the way, except now for computer sciences, earn the highest salaries by far in the humanities and sciences divisions of universities. They should be leading this.

So, SFI should take this role because it is not only the ultimate pragmatic problem; it's one of the ultimate complex-adaptive-system problems. And one of the things that has been missing, which has been said here either implicitly or explicitly, is that we need to bring all of these things together. It's not physics. It's not economics in its traditional, rather boring, sense, nor is it computer science. It's *all* of these. And SFI is one of the few places which is dedicated to that—it's really our stock in trade. So, that's the first point I want to emphasize.

Secondly, digging down a little deeper, one of the things that has astonished me that also has not turned up in this discussion is focusing a little more on growth. Because growth is underlying a lot of this and is either implicitly or explicitly incorporated within economics and finance. And that is that the only system we know that works, that has brought us all these wonders that we all participate in, the quality and standard of life, which we love and enjoy, is open-ended, exponential growth.

The question is, is that actually sustainable? And—I guess he was a maverick—Kenneth Boulding, the economist, when testifying in front of Congress, said, the only people that could believe that you can have open-ended exponential growth with finite resources are either idiots or economists. That's pretty much my own view, frankly. And so the next question is, how do we come to terms with that? We need to understand that much more deeply.

The Nobel Prize was given either last year or the year before for growth to my friend Paul Romer, who certainly doesn't address this question, which is fundamental not just to economics, but to the future of the planet. Associated with that is something that Eric mentioned. I'm also surprised—although I heard something this morning that may contradict this—that the whole conceptual framework of thermodynamics and system mechanics isn't actually integral to economics.

Indeed, I once did a little exercise: I took about half a dozen economics books, the big fat ones like Samuelson's, and so on, and looked up in the index: do the words "energy," "entropy," or "thermodynamics" ever occur? Not once in any of them. Energy! You can't even have a f——king dream at night without energy.

[LAUGHTER AND APPLAUSE]

Energy, and it's not there. I mean, it's ridiculous. Frankly, none of this would matter but for one thing, and this is the other reason that it's outrageous that economists are not taking a lead role. They are the heart of the academic community that politicians and practitioners, for all the wrong reasons, actually occasionally listen to, and they help form policy. So they have a special role that they should be taking. I've run out of time. I'll stop, but I was told to be provocative.

W. TRACY Well, Geoffrey, thank you for both being provocative and staying on time. That might be a first.

S. PAGE Only for the latter, though.

W. TRACY True. I actually want to dig into something that Geoffrey said and maybe to motivate it a little bit. You know, because the Berlin Wall fell thirty years ago today. There was this probably apocryphal story of Stalin trying to impress a foreign diplomat with a military parade. They have missiles and tanks and then they have this group of middle-aged people in ill-fitting clothes, and the foreign diplomat says, "What's that?" And he responded, "Oh, those are our economists; they can kill far more people than all the other weapons you've seen." Maybe with that as prologue, what idea from economics—traditional economics or even modern complexity economics, if you're feeling brave—do you think has done the most damage or has the most potential to do damage to the way that we make policies and actually apply the science and make decisions?

R. SETHI Rational expectations. I'll tell you why—for two reasons. Firstly, because it embodies equilibrium assumptions of the kind that Brian Arthur elaborated on. The assumption itself basically says that the subjective probability distributions that we hold in our minds about the future values of economic variables are such that our behavior then generates objective distributions that match those beliefs. So it's massive economy-wide coordination on equilibrium behavior. The problem with that, aside from its lack of realism, is that it's marketed as a behavioral assumption and not as an equilibrium hypothesis.

I just want to quote Herbert Simon, since a lot of people have mentioned him. This is what he said about rational expectations:

> *If awards were made for outrageous audaciousness in the coining of terms, certainly the author of "natural level of unemployment" would deserve one, as would Jack Muth, the author of "rational expectations." The virtue of such terms is that they win the argument instantly by taking the breath away from would-be disputants whose very skepticism now accuses them of unnaturalness or irrationality, as the case may be.*[2]

S. PAGE Yeah. I think it's the fact that we don't consider the spillovers. This gets back to some of the stuff I think Jessica was hinting at, that we only look at certain dimensions. So it's sort of like, is this efficient? But not what happens to social networks, what happens to the environment? Anything not priced isn't considered in the model. So you go talk to government agencies and you ask, why do economists have so much influence? They'll say, well, if you talk to an anthropologist, they just basically tell you, it's complex. And the economists have these simple models where they actually give you a prescription for how to act. They give you data, they can explain it in simple terms and it does have this sort of scientific sheen to it.

And so the question is, do you try and complement that with other sets of models? I think that's part of what the Santa Fe agenda has been, right? To think about the implications of these sorts of choices. We've talked a little bit about Silicon Valley and stuff, but you can imagine

2 Simon, H. A. 1997. *Models of Bounded Rationality, Volume 3: Empirically Grounded Economic Reason.* Cambridge, MA: MIT Press, p. 365, note 10.

I can construct a business, right? I can have a model and it can prove that I'm going to make money. But the spillovers from that could be pernicious in large ways and even small ways.

In Ann Arbor, there's a movement against ordering your Starbucks coffee and then just picking it up, because you're not actually interacting with a human then. It's dehumanizing the workers at Starbucks, as opposed to a locally owned cafe where I actually interact with the person. I might get to know them. We may develop a bit of a friendship as opposed to, you know, just punching something on an app and grabbing it. Matt could draw the social networks for us. But the thing is, there are implications for those social networks when you think about our willingness to accept inequality. Eric laid out this notion that we are social animals, and that we're not surely self-interested, but the reality is if we allow technologies and policies to isolate us—you know, if my only friend is Alexa—those preferences could go away.

C. M. CAPRA I would agree with Rajiv that the biggest problem has been rational expectations. They have had an enormous influence in macroeconomics. This might sound unfair, but to me macroeconomics is in crisis. And I will say that the right approach to save macroeconomics is what Blake is doing, for instance, with agent-based modeling, but also other people like John Duffy at the University of California, Irvine. They're looking at experimentation to better understand dynamic processes in macroeconomies. I mean, there is some effort going on, but it has been incredibly difficult to change the mindset. If you think about it, it's very inefficient, because a lot of these people are extremely smart. Why are they wasting all these brains, time, and energy on something that is obviously wrong?

I have a hypothesis and it has to do with the way in which economics departments are organized. Economics departments are organized more like philosophy or English than science. We are not really a science. So, the pace of change of ideas is very slow. I did some work in neo-economics. I was at Emory University working on neo-economics for ten years. The pace of change was so fast. You can't keep up with the knowledge, in a way, but the sciences devised a structure of their organization to generate a flow that comes from the postdocs, the grad students, people visiting, and so on and so forth—interdisciplinary research that allows people to update the methods and ideas more quickly. And so I think that's the problem the economics departments are facing: the way in which they are organized.

Just to give you an example, if you call out for papers, that's not good for tenure. It's really hard to do a good job if you're not collaborating. When I was up for tenure, I had to have a file for economists and a file for more interdisciplinary people. I had to produce in traditional economics and interdisciplinary at the same time. So just the interdisciplinary will not count. I think that's a consequence of the way in which we have organized. It is changing a little bit. You see a lot more postdocs, especially thanks to behavioral economics and experimental economics and data science. We see PhD students go into data-science postdocs and so on.

S. PAGE Let me just push back on this bit—and, Blake, bail me out here if I screw this up. If you go to the New York Fed, which I revisit once a year, they will say their models work really well. One way to think about how they think about their models is how I explain it to undergraduates: imagine a dog chasing a rabbit in your backyard. So the rabbit is a little *r*—think of that as this random thing and the dog, *D*, is deterministic. So a stochastic, dynamic, general equilibrium model—just let that go, all the paradoxes and consistencies and the notion of a dynamic stochastic general equilibrium. The idea is that the rabbit moves and then the dog deterministically goes to the rabbit. The way they test their models in terms of whether they are working is with a shock to the system. Then, does a rational-expectations approach, a real business-like approach, predict the dog moving towards the rabbit?

The answer is . . . it sort of doesn't do that bad. But they're not modeling the rabbit, right? The rabbit isn't random, like the 2009 stock market crash. It's not like a lightning strike in a collapse; the system itself is creating the rabbit. So you can think of the economy as a series of shocks moving to an equilibrium, shock, moving to an equilibrium. For some things, that's the appropriate way to model, but then there are other things that are much more complex adaptive systems and we have to model the rabbit. If you ask macroeconomists—if we had five leading

SEEKING A ROBUST MODEL OF DISEQUILIBRIUM

Rajiv Sethi

Modern economies are characterized by path dependence:[1] seemingly insignificant early decisions or minor random shocks can have cumulative effects that become highly consequential over time and can change the course of history. This is also true of the study of economies: fortuitous exposure to an idea early in one's academic life can alter the course of a career. I can vouch for this through personal experience.

One of my favorite activities as an impecunious graduate student was browsing through the shelves in the basement of the iconic Strand bookstore in New York City, where review copies of recently published books are still sold at deep discounts. It was there that I happened to chance upon a slim volume called *The Economy as an Evolving Complex System*, edited by Philip Anderson, Kenneth Arrow, and David Pines, based on a workshop held at SFI.

The chapters in this book were all concerned in one way or another with the economy, but their vision contrasted starkly with that found in standard textbooks. It was as if these chapters were describing an entirely different and vastly more interesting object. Where the textbooks featured mutually consistent plans unfolding along equilibrium paths, this volume opened my eyes to genetic algorithms, rugged landscapes, disequilibrium dynamics, and, more generally, processes of adjustment in complex environments.

There are many famous and familiar names among the contributors to this book, but the chapter that had the most enduring impact on me was written by someone rarely mentioned today: the Brazilian economist Mario Henrique Simonsen. His chapter was titled "Rational Expectations, Game Theory, and Inflationary Inertia."

Simonsen was a former finance minister of Brazil and very familiar with the practical difficulties faced by policymakers. His chapter involved mathematical reasoning applied to a very real policy problem: how to deal with hyperinflation. He started with an abstract question: when might Nash equilibrium be a predictively useful hypothesis in game theory? He classified games into those in which it could reasonably be assumed that players would coordinate on equilibrium strategies and those in which such coordination would be unlikely; he called them A and B games. This was where I first encountered what today would be called a beauty contest or guessing game; in this chapter it was called half-the-average.

Simonsen argued that a Nash equilibrium would not be predictive in this game, and he was correct. About seven years later, Rosemary Nagel published a classic experimental paper in the *American Economic Review*[2] that established this decisively, and set in motion a large and still-vibrant literature on alternative solution concepts in the theory of games.

But Simonsen's concern wasn't just with abstract theorizing; he was concerned with more practical issues. There was a debate raging at the time between those who favored gradualism and those who argued for shock therapy in addressing high inflation. Shock therapy involved a drastic cut in the rate of growth of the money supply coupled with a public announcement to shift expectations. The argument was

---

1 Arthur, W. B. 1994. *Increasing Returns and Path Dependence in the Economy.* University of Michigan Press.

2 Nagel, R. 1995. "Unraveling in Guessing Games: An Experimental Study." *The American Economic Review* 85 (5): 1313–326. http://www.jstor.org/stable/2950991.

that inflationary expectations in the economy would jump down discontinuously and actual inflation would drop without the need for a sharp rise in interest rates or mass unemployment.

This argument depends on complete and uncritical acceptance of the rational expectations hypothesis, which states that the subjective probability distributions held by agents in the economy are self-fulfilling, in the sense that they match the objective distribution to which they give rise. As Herbert Simon pointed out,[3] this hypothesis does not "correspond to any classical criterion of rationality," and the label accordingly provides it with "rather unwarranted legitimation." It is an equilibrium hypothesis asserting the mutual consistency of individual plans rather than a behavioral assumption akin to rational choice in consumption or production decisions.

Simonsen's point was that there are situations in which the consistency of beliefs assumed by equilibrium theory can lead one astray in dealing with important practical problems, and that this necessitated an explicit consideration of disequilibrium dynamics.

The point remains valid but largely neglected to this day. Economic theory continues to lean heavily on a notion of equilibrium that requires massive coordination and consistent expectations involving millions of economic agents without specifying a plausible process that could conceivably allow such coordination to arise. In fact, as far as the graduate school curriculum and the content of the leading journals are concerned, economics has moved towards even greater reliance on equilibrium methods and even greater resistance to nonequilibrium reasoning.

Why, despite some obvious problems with the equilibrium approach, have explicit models of disequilibrium dynamics failed to take hold? A cynical reason is that the profession is stuck at a local peak on a rugged landscape, with no incentives for anyone to explore distant terrain. But there is a more valid reason that is worth contemplating. Disequilibrium models in general—and agent-based computational models in particular—are too easy to construct and too hard to evaluate. There is really no accepted methodology that allows an editor or referee to clearly and convincingly distinguish a robust model from a fragile one.

Missing is a canonical model of disequilibrium dynamics that can serve as an exemplar for graduate students and young researchers, in much the same way a classic paper by George Akerlof opened the doors to a generation of work in the economics of information.[4]

It is worth recalling that Akerlof's paper was rejected by three journals before eventual publication.[5] Somewhere, perhaps yet to be written, is a paper that will do for the economics of complexity what Akerlof's paper did for the economics of information. When that sees the light of day, the floodgates will open, and the promise of *The Economy as an Evolving Complex System* will finally start to be fulfilled.

Further Reading

Sethi, R., and M. Yildiz. "Communication with Unknown Perspectives." *Econometrica* 84, no. 6 (2016): 2029-069. http://www.jstor.org/stable/44155357.

---

3 Simon, H. A. 1978. "Rationality as Process and as Product of Thought." *American Economic Review* 68 (2): 1–16.

4 Akerlof, G. A. 1970. "The Market for 'Lemons': Quality Uncertainty and the Market Mechanism." *The Quarterly Journal of Economics 84* (3): 488–500.

5 Gans, J. S. and G. B. Shepherd. 1994. "How are the Mighty Fallen: Rejected Classic Articles by Leading Economists." *Journal of Economic Perspectives* 8 (1): 165–179.

macroeconomists up here—they would not say we're terrible what we do. Because the metrics they're applying to their own work are a particular lens and, viewed through that lens, they're not doing that badly.

R. SETHI So, we've heard, especially from Brian Arthur, about the book that came out thirty years ago, *The Economy as an Evolving Complex System*. Beautiful thing. I discovered it as an undergraduate; it changed my entire career. It was hugely influential. And we've heard about the contributions of people like John Holland and Stuart Kauffman.

I would like to mention the contribution of someone we don't hear much about, who has a chapter in that book that for me was by far the most influential. It was former finance minister of Brazil, Mário Henrique Simonsen. He wrote a chapter called "Rational Expectations, Game Theory, and Inflationary Inertia." And this where I want to push back on what you said and really support what Mónica said. He was a practical policymaker. He was fond of abstract thinking and the chapter has got a lot of abstract thinking.

He basically starts off with a question: in what kind of games would we expect Nash equilibrium to predict well, and where would we expect it to predict badly? And he divided them into what he calls A and B games, and then B into B_1 and B_2. Part of the B games are what today we would call beauty contests or guessing games. He called it half-the-average game, and he said that this is not a situation in which the Nash equilibrium prediction is going to work.

Seven years later, Rosemarie Nagel published a classic paper called "Unraveling in Guessing Games" in the *American Economics Review* and it proved him completely right. Nash equilibrium is hopeless in this game, and anyone who plays the Nash equilibrium strategy has very low payoffs compared to those who reason in a different way. Why was Mário Henrique Simonsen interested in this? Because there was a big debate at that time about gradualism versus shock therapy.

He's the finance minister from Brazil. In bringing down hyperinflation, he's dealing with this very practical, very important problem. Some of the rational-expectations theorists at the time were saying you can reduce inflation suddenly and quickly by making a public announcement that you're going to contract the rate of growth of the money supply. The entire public will change their inflationary expectations discontinuously. You don't have to have mass unemployment and you don't have to have massive increases in interest rates. This is exactly the opposite of what the Fed did in the Volcker years; they raised interest rates, they had large unemployment, and they brought inflation down. This is gradualism. Folks were saying, "No, you don't need that."

Mário Henrique Simonsen's point in this chapter is that the rational-expectations hypothesis relies too much on equilibrium reasoning and that it's leading you astray. And he was totally right about that. I would like every single graduate student in this country to read that chapter in that volume. The reason why the Fed does quite well is by ignoring these kinds of predictions. Right now there's a debate among economists about the neo-Fisherian principle, which states that if you're worried about deflation, you should raise interest rates instead of lowering rates towards the zero lower bound. The economy will assume that you are expecting higher inflation and will jump to this new equilibrium. It's absurd. You know, the reason why the Fed actually does well is it knows when to ignore this kind of stuff.

. . . is thinking in terms of complex adaptive systems just adding a few nuts and bolts to what's there, or is it revamping the whole thing? There are some major pieces missing in economics that have biological analogs that need to be put into economic thinking.

W. TRACY Geoffrey, do you want to get in on any of this?

G. WEST Not on this. Because this is down in the weeds of economics. And I only very superficially know about economics at this level. But I was ruminating as they were talking that this session is called "Economics as an Organism" or "Organismal Economics." So far that hasn't come up, and I was wondering whether we want to talk a little bit about that.

One of the themes that has come up in the discussions is indeed the whole question about to what extent you could put it as, to what extent is economics, since it's a complex adaptive system, also acting in a biological fashion? That is in contrast to the ways that have been talked about because of the questions of adaptation, evolution, and so forth. Something that also has not come up in this so far other than, again, implicitly is that biological systems operate—as in fact any complex system does, ultimately—by the interface between energy and information.

In fact, both of them can be thought of in terms of energy networks, the distribution of energy to the multiple components of the system and the processing of information and distribution of information. It is sort of interesting that economics doesn't seem to think in those terms. It's just beginning to touch the surface, and we heard Matt's talk yesterday of incorporating networks into it and it's again surprising that even though networks have been part of the social sciences for an extremely long time, they never really penetrated economics, even though the whole process is a network process.

And that network process, just to repeat myself, is a process of distributing energy, which maybe loosely could be thought of as assets and goods and so forth, but also information and information theory. So, that's sort of the big picture. Digging down a bit deeper is another piece that doesn't get a lot of press within economics, and that's something I happened to get interested in, which seems to be central to economics. That is that economics happens primarily in cities. And the most amazing thing about human beings is that's where we live and it is by bringing us together and getting those networks to be denser that leads to all the marvelous things we have.

And somehow that isn't part of economic thinking. There is open economics, of course, which actually only touches on that. So, I'm just sort of ruminating on that as I listen to this conversation, which I am not part of, actually. But, also, I'm trying to think in terms of the Santa Fe Institute and thinking in terms of complex adaptive systems. I disagree with some things Brian said—I think he's over there. But I love the idea, this question: is thinking in terms of complex adaptive systems just adding a few nuts and bolts to what's there, or is it revamping the whole thing? There are some major pieces missing in economics that have biological analogs that need to be put into economic thinking. One of them is the idea that we have a species and we have taxonomic groups. We have those in economics, too, except you don't talk about them. They're called cities and companies. We do talk about companies, but these always play subdominant roles.

One last thing that is also missing is something that's to some extent missing in biology, too. But it's certainly missing in economic thinking, and that is time. I mean, literally, time. That is, that time is speeding up.

Time was very different when Mr. Smith wrote his book or when John Keynes wrote his book. Time is completely different now. It's speeded up and it's speeding up all the bloody time and it goes back, closing the loop, back to the sustainability of the planet. We live on an accelerating treadmill and if you don't put that into your equations, into your computation, into your conceptual framework, you're doomed not to be able to describe the system long term.

Short term, it's okay—that's why these models of short-term projections into one or two years might even work. So, there's a huge issue there and a much bigger issue that as time speeds up and things get faster, your body and your brain barely has changed. That is, basically, we're the same thing as we were a hundred thousand years ago, and the marvel is that we've been able to adapt to the typewriter, to the telephone, that we're able to get on steam engines and go faster than four miles an hour. And now we can pick up iPhones, but it's going to get faster and faster. Somehow this should be integrated into economic thinking.

S. PAGE Right. Let me jump on this. So, Josh was speaking about this, this notion of thinking of actors as sort of being hyper rational and jumping to equilibrium versus adapting is kind of at the core of the distinction of these two ways of thinking. Let me just explain the Nagel game for a second. So, in the Rosemarie Nagel game, everybody makes a prediction. Whoever's closest to two-thirds of the average wins. If everybody guessed fifty, whoever's closest to thirty-three would win. This means that in equilibrium everybody should be at zero.

When I was teaching at Caltech, Rosemarie gave a talk and Charlie Plott said, "Let's play it for fifty bucks." It was a bunch of game theorists, so we all thought this was silly, but in the room was the husband of one of our grad students who made stained glass. So then we're all sitting there, everybody's going to say zero. And they're like, "Oh, but what's the stained-glass dude going to do?

[LAUGHTER]

And then I had to think, what does Rajiv think the stained-glass dude is going to do? And then I think, what does Rajiv think *I* think the stained-glass dude is going to do? And it ended up the average guess was twenty.

R. SETHI Yup. That's usually the winning . . .

S. PAGE Yeah. You get completely unraveled. Matt Jackson had talked about this before. There are some things in the economy that we call *dominant-strategy mechanisms*, where everybody should just tell the truth and you'd imagine sort of rational economic logic to drive things the right way. So if there is a shock to the economy and oil prices go up or oil prices go down, we can kind of expect expectations to work in the rational model to kind of play that thing out. In a hyperinflation that's not going to work.

R. SETHI Yeah.

We live on an accelerating treadmill and if you don't put that into your equations, into your computation, into your conceptual framework, you're doomed not to be able to describe the system long term.

S. PAGE But extending that, and I think Geoff raising this really interesting point that came up in a lot of the earlier talks is that we're not modeling Agent_Zero. Geoff's work is so fantastic in this space. Cities, just like ecosystems, are more diverse as they get bigger. Why is that? Well, if you think about it from a biological perspective, there's just more diversity carrying capacity, right? Like if I'm interested in

European soccer and I live in a town of 500 people, there's just no one to talk to about that. You can actually sustain an interest in something if there are more people there.

And so what you get then is cities are more interesting *because* they're larger. But then if you look at all the literature on where does innovation come from in science, where does innovation come from in the economy, it comes—a lot of it—from recombining ideas. Eric's written on this at length; Brian has looked at this even more. Well, then, it's more likely to happen in cities. But the standard neoclassical model doesn't help us. I think the biological model does.

G. WEST Exactly. Thank you.

W. TRACY I don't know if anyone else wants to comment on what Geoffrey said. I know, Mónica, you do a lot of work in related areas, in particular integrating networks into economics.

C. M. CAPRA Yeah. So, I think the previous panel talk about what the main purpose historically for economics was: to understand efficient resource allocation. And so initially thought of it happening through market interactions. So, a lot of economic research has been on understanding markets, right? But it's true that a lot of economic activity and society activity happens within networks and that's kind of like the work that Matt has done in contributing to that idea that networks actually play a super-important role.

I think that economists have been limited with respect to the availability of data. Remember, most of history we have been dealing with not enough data, and it's only recent that we have all these data. I mean, we can understand the interactions because we can have GIS checking where it is that the chips are going, and we can have phones and other things that will give us data on understanding how these interactions within networks happen. So I do think that economists are adjusting, but, in a way, it is slow and slower than it should be. In fact, I think it's because of what I just said, the way in which the discipline itself is organized.

W. TRACY We have about fifteen minutes left. So, I do want to open it up to the floor for some questions—Ray?

RAY IWANOWSKI[3] I'm going to do something pretty dangerous here. I'm going to try to defend, in some form, classical economics—as applied to finance and asset pricing theory. If you think about a few of the prevailing models in those fields, they were derived from first principles—expected utility maximization.

These models, developed over fifty years ago, started with some assumptions and, upon derivation, led to a few conclusions with testable implications. One, the market-efficiency hypothesis, concludes that market prices quickly and effectively reflect all relevant available information. Another is the two-fund separation theorem, which leads to the Capital Asset Pricing Model (CAPM). A key implication of this model is that it is optimal to hold some combination of the market portfolio and cash.

Almost immediately it was known that these models were based on somewhat unrealistic assumptions and were almost certainly false to some degree. It has been empirically demonstrated that markets are clearly not "perfectly" efficient. And many have made the case that certain portfolio allocations may be superior to a linear combination of the market portfolio and cash, both ex-ante and ex-post. Over the years, the finance literature has identified several hundred anomalies that represent empirical violations of either market efficiency and/or the CAPM. And researchers have identified behavioral biases as reasons why there might be violations of the theory. Armed with this evidence and understanding of the sources of the atheoretical findings, some of us who are investors may trade and profit from these "anomalies."

But, as a prescription to investors, the theories may be reasonable straw-man starting points. Certainly, starting with the premise that markets are fairly efficient implies that investors should not try to spend too much time, effort, or transactions costs and fees trying to beat the market. This still seems like sound advice to most investors that held up well over the past fifty-plus years. And asset allocations of holding some version of the market cap–weighted portfolio with some cash (the weight between the two adjusted based on risk tolerance) implies buying and holding index funds. This strategy has produced superior returns

---

3 SFI ACtioN member and Managing Principal/CIO, SECOR Asset Management.

to most alternative strategies over the years, especially after accounting for fees and transactions costs.

And I think Josh Epstein made a great point—in the current state of economics, there's does not appear to be a viable alternative theory that incorporates these critiques of the classical theory that were derived from first principles. It is fine to reject the theory by saying, for example, that agents who price assets are not rational but are subject to behavioral biases. However, there does not appear to be a plausible theory that leads to better advice and prescriptions from the implications that arise from market efficiency and CAPM—i.e., buy/hold index funds and do not try to beat/time the market. There are empirics that better explain the data but not a theory that arises from these findings that lead to more optimal behavior.

I just started getting involved with SFI, learning about complex economics, reading about it and coming to these conferences. The idea of the market as a complex adaptive system has amazing appeal to me. The agent-based modeling that we discussed yesterday morning was very cool. But I will say that there doesn't seem to be a set of deliverables that say, "Okay, classical economics is clearly flawed. So, instead of holding the market portfolio, do this." One implication that has been highlighted is that classical economics does not properly account for crashes. Okay, so we now have models that can do that. But what should market prices do? Should crash-susceptible securities be priced at lower prices? Does complex economics suggest a portfolio theory that implies that you need to hedge against such crashes? If so, how?

I still haven't seen an alternative CAPM that describes how the expected return on assets should be priced—for example, adjustments/deviations from the CAPM—that account for the recognition that markets are complex adaptive systems. Until such alternatives are suggested and empirically validated, it does not represent a viable improvement over the classical models that were accepted fifty years ago. Can we use complex economics to identify ahead of time when disequilibrium is going to come, such that we can alter our strategies or allocations to properly reflect complexity?

It just seems like there is a lot of criticism of some of the basis of classical economics—such as assumptions based on the concept of

THE EQUILIBRIST & COMPLEXOLOGIST CAN BE FRIENDS

Scott E. Page

A complex system's framework makes a certain set of assumptions about what constitutes the economy and the world. The economy, for instance, consists of diverse actors, who pursue objectives by adapting strategies based on feedback. The success of their actions depends on multiple network positions—the attributes of their products, services, or skills; their physical location; and their competitors and cooperators. The aggregate—something complex, unpredictable, perpetually novel—emerges from these micro-level interactions.

The neoclassical economic framework assumes a bit less heterogeneity and that the actors are rational, that is, utility maximizing, subject to informational and incentive constraints. In most instances, given that rational people close the cognitive loop (I know that you know that I know), these assumptions produce equilibria.

Both frameworks rely on models. Or, to borrow the Austrian economist Ludwig von Mises's imagery, both cathedrals have been built brick by brick: a no-trade theorem here, a rugged landscape there. These models improve our ability to reason, explain, design, communicate, act, predict, and explore (forming the superhero acronym REDCAPE).

The high priests in these cathedrals often position these two frames against one another. If I had a piece of turquoise for every Santa Fe talk that set fire to the straw man of *Homo economicus* and a piece of silver for every time I heard an economist lay waste to those ridiculous econo-physicists, I could open a rather large concho belt store.

Yet, if we look around us, we see some parts of the economy as best described by equilibrium models—commodity production jumps to mind. And we see other parts that appear far more complex—like commodity prices. To wit, at the beginning of the COVID-19 pandemic in April of 2020, cheese prices plummeted by almost 50 percent. In June, they rose 150 percent.

For this reason, as the pandemic unfolded, from what I could infer, most wise investors were of two minds: one equilibrium, one complex. The equilibrium frame told them that lower demand would reduce the price of crude oil. The complexity frame said, "You know, the price might even go negative."

If I had a piece of turquoise for every Santa Fe talk that set fire to the straw man of *Homo economicus* and a piece of silver for every time I heard an economist lay waste to those ridiculous econo-physicists, I could open a rather large concho belt store.

The value of both models extends beyond prediction. Opening up our REDCAPEs, if we consider any policy choice—be it economic or political, or the design of a program or strategy, we see why this is true. Is the current international political system better understood as a chess match between the US and China, with pesky Russia occasionally removing a piece from the board, or as emerging from networked actions of governments, NGOs, and corporations?

Yes, as Anne-Marie Slaughter would say, it is.

Of course, in contexts with less sophisticated actors and more complexity, the complexity frame usually wins. The COVID pandemic meets both assumptions. Models in which people choose equilibrium levels of social distancing and mask wearing fared poorly compared to models of networked behavioral updating and norms. That said, solving for equilibrium behavior does provide counterintuitive insights.

Attaining that methodological consilience will require acceptance from Hyde Park and Cambridge and a touch of humility from the land of enchantment. We might heed the mid-career lyrics of Oscar Hammerstein: "I don't say I'm better than anybody else / But I'll be danged if I ain't just as good!"

Further Reading

Page, S. E. 2018. *The Model Thinker,* Basic Books, NY, NY.

rational expectations—but no superior alternative offered beside the recognition that markets are complex. Most people that I know who still find classical theories useful don't believe that all human beings are rational. They simply accept that irrationalities may not matter or may lead to second-order effects. Or they may believe that these effects are significant but still find the theory as a useful straw-man starting point from which deviations can be measured. So, I would say that if you want to criticize these models and the concepts that arise from them, an alternative should be offered that has been proven to explain the data better and leads to better advice to the investment community and market participants.

R. SETHI I'd be happy to respond to that. Let me just focus on the point about markets incorporating very quickly all available information. There are two aspects of the efficient markets hypothesis; one is quite obviously true and the other I think is problematic. So, one is that the markets are hard to predict. That when there's a news event they move quickly and they'll do something. And they'll move in the direction in which the information tells you to move. And it's very hard to beat the markets, partly because there's a lot of competition for alpha and so on and so forth. But how do you interpret this? Markets being invincible is not the same thing as markets being efficient. And there's an alternative interpretation of what happens when there's news in market.

There's a wonderful essay by John Kay,[4] actually, and I wish I could remember the title of it, where he basically describes the price of an asset as being determined by a clash of narratives. It's not about the information; it's about the interpretation of information. There are stories. There are folks who believe that the price is too low, because they have in their mind the story that will cause the price to go up quite substantially in the future. There are people who have got a story in their head that the price is too high, that they should be really shorting this asset. And those people have maxed out on the risk limits. They can't take on any more exposure and so you have a stable sort of situation where these narratives are basically balancing each other out.

---

4 https://www.johnkay.com/2011/10/04/the-map-is-not-the-territory-an-essay-on-the-state-of-economics/

When you get a crash, it's when one of these narratives collapses, something happens where the story just cannot be sustained. There's a wonderful book that's just come out, by Bob Shiller, called *Narrative Economics*, where he discusses this interpretation of what happens with regard to markets. Now it's not as clean, it's not as formal, it's not as neat as the efficient-market hypothesis, but to me it just has a more realistic feel to it. And I think that pointing out that markets are hard to beat is not really direct evidence on the interpretation that they incorporate all available information. It's the interpretation of information is a really key part of asset pricing that's really missing from that perspective. That's my take.

> Markets being invincible is not the same thing as markets being efficient.

G. WEST I'll defend the classical economics. And I'm on dangerous ground because I don't know very much economics, but I think Brian stated it again last night. It is a beautiful theory, and it's inspired, I guess, by physics. It's beautiful because it pretty much sets out what the assumptions are. So as if you assume the following—some of which may well be wrong—these are the consequences. Then there's a whole body of work built up and, indeed, if the situation remains relatively stable then it's actually relatively predictive as I understand it. You know, it works pretty well. That's the way physics works. We make these models and we have very well-defined assumptions and we use those as points of departure for either elaborating on them or for making grand leaps.

But Josh Epstein this morning made the important point about physics models, that they are abstractions, just as the classical models of economics are. The best example is the famous one of Galileo, the mythical throwing objects off the leaning tower of Pisa and the discovery that they all fall at the same time. Which is certainly not true. But I once made the statement I got trashed for, for saying that if Galileo had been a biologist or an economist, instead of concluding from what he observed that everything falls at the same rate, abstracting that, he would've said, well, you know, if it's got a fuzzy surface, it will fall slower than if it's got a sharp point at the end, and doing the great taxonomy, writing tomes about the details of the great diversity of the ways in which different objects fall.

Which is exactly what does happen out there. The great genius of Galileo was to abstract to a situation that has never existed on this planet ever. It can't, in fact. And in that situation, all objects fall at the same rate. And beginning with that, we got Newton's laws and then we got thermodynamics and then we got Maxwell's equations and then we have iPhones and everything came from that kind of thinking. So I'm a great defender of what was done. The problem seems to have been that it stopped at that point, that there wasn't sort of an Einstein or a Schrödinger or whoever. I'm going to really get clobbered for this, but Samuelson and Ken Arrow and all these people weren't Einsteins or Schrödingers.

They're just very smart people who just added on to the same old stuff. So that's my provocative statement. What we're waiting for is for an Einstein to come along and say, "Look, guys, don't look at it this way. Yes, that was all great, marvelous. But if we say that the velocity of the stock market is the same everywhere in the universe or whatever, some idiotic statement, all of this follows and we understand that within the next eight to ten years there's going to be another great crash, and in fifteen to twenty years the whole bloody thing is going to go down." That's what we need. Something like that. So I am a great defender because it's in the right tradition of what I consider the way you should be doing science.

S. PAGE In some ways you're echoing . . . Roger Myerson, who won a Nobel Prize in economics, argues that, one, you want to start with the rational actor model because it's a reasonable benchmark.

G. WEST That's the point.

S. PAGE Myerson won the Nobel Prize for something called the *revenue equivalence theorem* that basically says if you have rational actors, it doesn't matter how you set up an auction, right? You could have highest bidder wins, second-highest bidder wins, increasing bids, closed bids—it doesn't matter, under some restrictive assumptions. But if, under the assumption of rationality, institutional design is irrelevant . . . Josh is peeling back papers ready to roll here.

[LAUGHTER]

The reason why is, once you relax the assumption of rationality, there's huge differences. And what's really funny is when I go to auctions, charity auctions, about fifty percent of the time, the auctioneer, I'll say, "I'm a Myerson student," the auctioneers say their goal in life is proving Myerson wrong. And what they do is they just get under David's skin and get him to act emotional and irrational and overbid something. So then it's David against Bill, and they raise it.

So the whole point is, they have their social preferences, there's a pecking order. You get emotion involved and Myerson's theorem is out the door. Right? Because people aren't rational and dispassionate. So I think I agree. The point is not to throw out the rational actor model; it's a good benchmark. From there we've got to ask, let's have other models.

G. WEST Exactly.

C. M. CAPRA Yeah, I completely agree with what he said. It's a useful benchmark, of course. We couldn't even think of how to design an experiment without that benchmark.

S. PAGE Right.

C. M. CAPRA And there's no way we're going to drop, you know, Newtonian physics. I don't think you're going to do that. That's not the point at all—it's actually to enrich the models that we have. But let me tell you something interesting from doing experiments with human subjects. Always we get between ten to fifteen percent of the subjects who are perfectly *Homo economicus*. It's really interesting. And then, if you come with this idea of heterogeneity . . .

S. PAGE And they become economists too.

C. M. CAPRA Those are the ones who decided to get a PhD in economics and they see the world differently, right? But we have this heterogeneity. Actually, I have a paper recently with Rosemarie—the beauty contest game. There are some people even in Amazon Mechanical Turk who find the natural equilibrium right there. About ten percent of the population seems to be just thinking that way. It's interesting. How is it that the different proportions of groups of people affect the outcomes? There's a paper also in acid market bubbles where, if you combine experienced subjects with inexperienced subjects, as long as you have about thirty percent of the subjects experienced, the bubble is going to disappear. These are some things that, again, we wouldn't know if it's a *Homo economicus* or not if we didn't have the initial benchmark model.

. . . in economics we've created simple models that often don't have verisimilitude. They do violence to reality.

W. TRACY Rajiv, unless you want to jump in, I want to open this exact topic up to some of the other people on the floor as well. Eric, did you have something you want to say here?

ERIC D. BEINHOCKER Well, yes. I was going to somewhat agree with Geoff and company, but also disagree—have a little philosophy of science debate. So, abstraction, simplification, and parsimony are good things. And models, one hundred percent agree with that. But there also has to be verisimilitude. You know, it has to actually capture something true about the world. And if Galileo's simplification had predicted his object would fly up rather than down, we would've said it's a pretty bad model despite its simple elegance. And so I think in economics we've created simple models that often don't have verisimilitude. They do violence to reality. Herb Simon we've quoted a lot today, but he had a famous quote that a lot of economic predictions don't even get the sign right half the time. You know, and at least you've got a fifty–fifty chance of that.

So, the key question is, can we create simplified abstractions that capture some essence of reality, don't stand in contrast to it, and are empirically falsifiable? Because if we don't, where economics has gone, we have a garbage in–garbage out problem. We have wrong assumptions that then lead through perfect logic to bad analyses.

G. WEST Eric, I took that for granted. I'm a physicist.

E. D. BEINHOCKER Yeah, but economists don't because economists use—this is Milton Friedman's fault, an as-if argument. If the model makes an as-if prediction . . . now, the example I give to my students is, the sun rises every morning as if Apollo was towing it in a golden chariot. It makes a perfectly good prediction but it's a lousy explanation. But economists don't recognize that.

W. TRACY So, I think we have time for one more question from the floor.

AUDIENCE MEMBER Scott, what was the Fed's response to the system?

S. PAGE Yeah, so, I think they're really working hard to think of it, how do they expand their hiring right? To hire sociologists, to hire historians, to hire anthropologists, to hire psychologists, but the Bank of England's had a lot—Andrew Haldane, who's chief economist there, has done a lot of work to try and bring in people from other disciplines. This is Kathy Phillips's work from Columbia—she's a Business School professor—just bringing psychologists into the room makes the economists all act differently.

[LAUGHTER]

In particular, they feel compelled to explain things to the psychologists, and then, while explaining it, they sort of realize, wait a minute, maybe there's a flaw there, right?

And so, one of the issues in this space is that, if you start taking things for granted, right, you don't question the assumptions. Putting people with other disciplinary training, with different background knowledge in the room, you know, it's uncomfortable but it causes the standard assumptions to be questioned and it just leads to better solutions. So the exercises that I've done at the New York Fed is just assign people roles. You're a historian, you know, you study transportation systems, you're a sociologist, you're an anthropologist, you're a psychologist, you know, you're a data scientist, discuss. This policy we're thinking about reducing inequality and it just enlarges the conversation.

It's in some ways better and worse than you think. It's better than you think, and the economists are fairly open-minded, but it's worse than you think when they form their own teams, people who went to Wharton will pick people who went to Wharton. People who went to

Harvard will pick people who went to Harvard. It's like the World Bank and the IMF. And so you end up with cases—we've had recent presidents, you look at their council of economic advisors, they all went to grad school together. That's a recipe for disaster, in my opinion. I mean, first of all, the fact that they're all economists, and second of all, that they're all trained by the exact same people in the exact same year, read the same papers, you know, it's probably not good, right?

So, what I try to do when I go in those places is to enlarge the frames and enlarge the discussions, like, do you have an environmental scientist in the room when you're discussing that policy? The answer typically is no. I think a lot of people in this room agree probably you should.

W. TRACY So, unfortunately, we have to cut the debate off there but let's thank all of our panelists again.

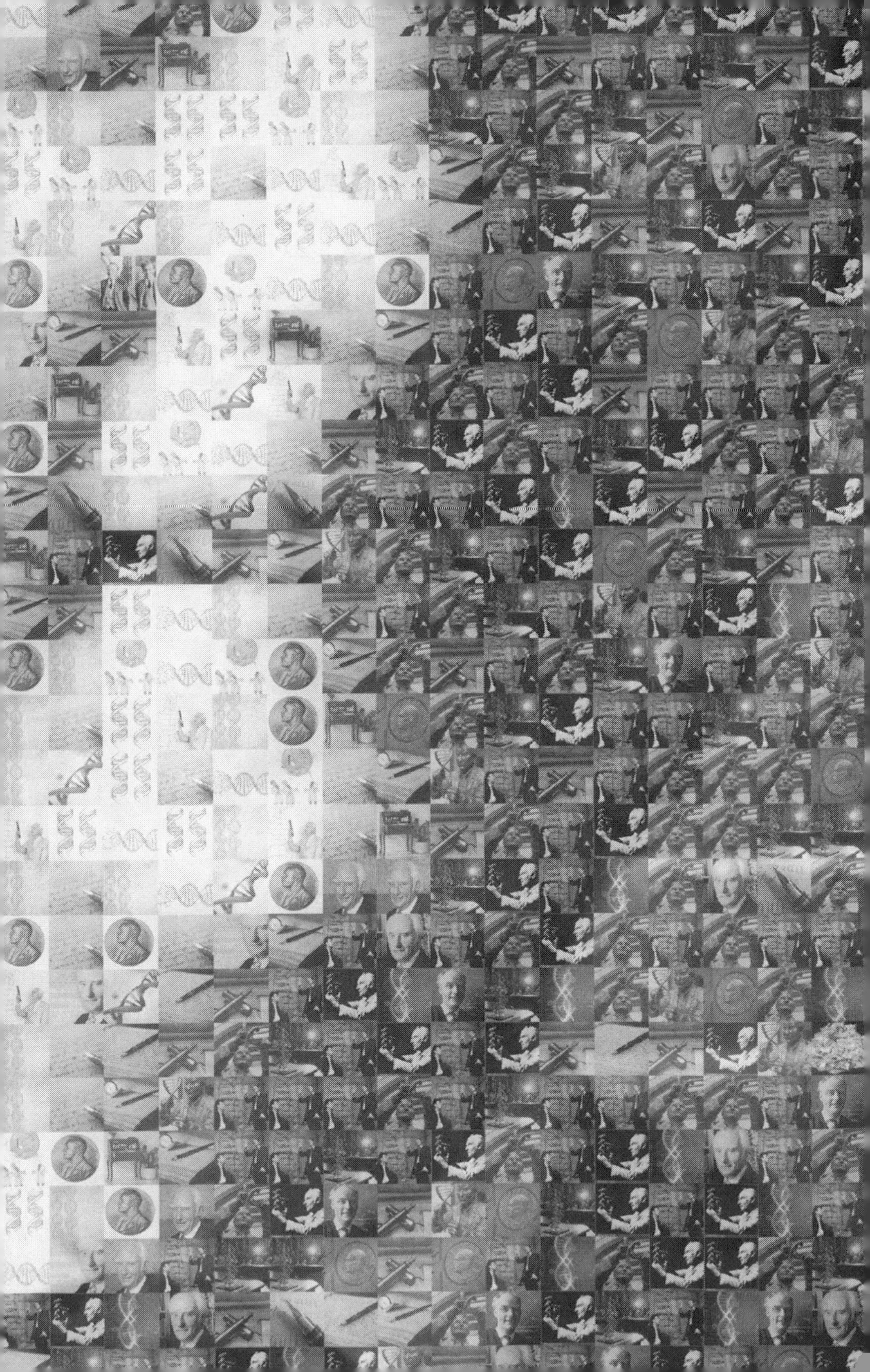

ECONOMIC ARCHITECTURES

[Panel Discussion]

Moderated by William Tracy,
featuring W. Brian Arthur, Eric D. Beinhocker,
Matthew O. Jackson & Allison Stanger

WILLIAM TRACY Our last academic panel of the afternoon is focused on economic architectures, which will include some discussions of networks, institutions, and structures within the economy. Obviously, these are all people that we heard from already in the program this weekend. I want to push each of you to think about the areas that really need to be developed next. Particularly, feel free to sort of transcend that positive economics/normative economics boundary as some of your talks did last time. So, not just what's needed for economics, but what we need to do in the way that we think about economic systems to better the world. Maybe, with that very small request, I'll start with you, Brian.

W. BRIAN ARTHUR Okay. I want to enlarge on something I said last night. Actually, it's come up in some of the other panels. In spite of a lot of what we know in economics, there still have been financial failures, economic failures, policy collapses, all kinds of things. The Euro was launched as a common currency quite a few years ago in Europe, and that turned out to be something of a disaster for Greece and Portugal and Italy, possibly also Ireland.

These things were not particularly well anticipated. When policies or policy systems are put in place, what economists tend to do is simulate and try to figure out what are going to be the consequences of that policy. That's a perfectly sensible thing to do.

OPPOSITE *Economic Curiosity: Francis Crick's Letter to His Son*

But if you think in terms of equilibrium, as I said, there's a subtle bias with all of us economists. If things are in equilibrium, there's no way to improve on that, otherwise it wouldn't be an equilibrium, so we don't think in terms of people gaming such systems.

With complexity economics, if we threw out this assumption of equilibrium, suddenly everything becomes a web of incentives that can be gamed or can be exploited in some way. Again, I don't want to be cynical, but I love this quotation from Janet Napolitano. In 2005, she was governor of Arizona, and people were bringing up the idea of a fifty-foot wall. So this is an older idea than the current administration. And she said, "Show me a fifty-foot wall and I will show you a fifty-one-foot ladder."

So, my question and challenge for economics, or for complexity economics, or even for SFI—it's not a huge challenge, but it's doable and it could be part of some program in the future—is, can we anticipate how such systems could be gamed, or played, or taken advantage of, and thereby collapse?

This is something medical professionals understand extremely well. In virology, in any sort of disease control, certain other organisms can get into the system, take it over, and cause severe difficulties, if not total collapse and death.

How would we get systems to do that? These would be intelligent systems, systems that can learn. A trivial example I thought of was, "Well, I could set up a model of fifty-foot walls. And then, trivially, I could feed it the idea that people could wheel up ninety-two-foot ladders or thirty-nine-foot ladders, or maybe fifty-two-foot ladders. And that would converge and discover that fifty-one-foot ladders would do the job just about right."

Now that's cheating. It's trivial because the whole qualitative idea's been thought of. So my question, I want to put it to you—I thought a lot about this. I've written a paper called "All Systems will be Gamed."[1] This was shortly after the collapse of 2008/2009.

But could we do better? Could we imagine an agent-based system or some kind of strategic counter-system that takes a given policy and

---

1 Arthur, W. B. 2014. "All Systems will be Gamed: Exploitive Behavior in Economic and Social Systems." SFI Working Paper 2014-06-016. http://tuvalu.santafe.edu/~wbarthur/Papers/All%20Systems%20Gamed.pdf

tries to discover how to game it? Such a system would have to ponder, meaning it would have to be imaginatively creative. How could I create some policy that would exploit this given system?

I don't think it's out of sight. I think it would need a lot of general knowledge. So, we're talking maybe about a form of general artificial intelligence where there'd be deep knowledge of how said systems have collapsed or been gamed or exploited in the past.

> . . . can we anticipate how such systems could be gamed, or played, or taken advantage of, and thereby collapse?

I want to point out that, when the US Navy or some other government body issues a new encryption method, then it puts that encryption method out to beta testers, mathematicians, and public, and invites them to see if they can crack it. If they can't—if nobody can crack it—it's viewed as a pretty good system.

So, my question in general to SFI, to all the really super-smart people here in the audience is, what would it mean to construct a clever, possibly agent-based, intelligent system that manages to figure out how policies or institutions can be gamed, exploited?

If we can do that, I think this would be a major way forward in saying we can have a much healthier economic system. We've done this with respect to earthquakes, with respect to aircraft failure, and many other ways. Time we did it in economics.

ERIC D. BEINHOCKER Thanks, Brian. So, I thought for economic structures I'd pick the uncontroversial topic of market versus states—asking what that might look like from a complex-systems perspective. And I'll start with two caveats. First, before anyone in the audience accuses me of being a wild-eyed socialist, I spent twenty years as a practicing capitalist before going back to academia. And my co-author in this work was the first investor in Amazon. So he's a rather successful capitalist.

What we're advocating is something different than our current form of capitalism, and not socialism. The second thing is, this is a personal perspective. But I think it's a big topic that SFI and this community needs to start exploring more.

So, first, in the neoclassical framework, we've had markets and states positioned as something in opposition to each other. And that markets are kind of a default setting for society, and they're efficient at allocating resources, but other institutions—notably governments—are considered inefficient. There are basically two arguments that support this.

One is the neoclassical rational-choice-equilibrium argument that markets automatically come to the Pareto optimal equilibrium for society. This was Ken Arrow and Debreu's great work.

The second is more out of the Hayekian tradition, that markets are efficient at processing distributed information to help coordinate activity in the economy. But both these views agreed that governments are less efficient than markets excepting a few specific conditions where you have market failures and public goods. They also both accepted that sometimes other noneconomic social considerations would require state intervention.

But—and this was the big but—there was always a big trade-off, which was the name of a famous quote by Arthur Okun in the '70s. Between economic equality and social issues and economic efficiency and growth. Now this kind of big trade-off has framed our politics since, where the left sees more market failures and values more the social issues, and the right sees fewer market failures and values more of the growth and opportunity side of things.

But, you know, despite the huge noise and tensions in our politics, they were both kind of agreeing on the same core theory. I would ask, though, does complexity give us a different core theory that might create different disagreements and different debates?

And I would put forward that markets and states really should be viewed as an ecology, a kind of interlinked, interdependent complex adaptive system. Asking whether you want more markets or more states makes as much sense as asking, "Do you want more plants or animals?" Right? It's a kind of nonsensical question. They're both part of the same ecosystem. Rather, the real question is, how do you get a healthy ecology?

BRINGING ECONOMICS BACK INTO SOCIETY

Eric D. Beinhocker

It's a system. It's an ecology. It's a network. It's made out of humans. These were some of the Economic Architectures panel's themes in providing a modern, scientifically grounded description of what the economy is. From a distance, this way of looking at the economy might seem almost trivially obvious, particularly to non-economists. But it is not how economists have looked at things over the past century. Instead the broad framework has been that of rational, self-interested, individualistic agents interacting in efficient markets, with non-market institutions, notably states, playing a secondary role. This way of thinking has had consequences; it has framed our politics, our policy debates, and how we are grappling (or failing to grapple) with issues like climate change, economic inequality, and the advance of AI.

All of the big challenges we are facing—including of course COVID-19 which hit shortly after the conference—are systemic challenges. Yet, ever since Lionel Robbins in the 1930s, economists have defined "the economic problem" as how we efficiently allocate finite means to infinite potential ends, and the answer to this has been "markets." There is of course some truth in that view and markets play a key role in all economically and socially successful societies. But this focus on efficiency, allocation, and markets has too often blinded us to the bigger problems. As I note in my remarks, the real "economic problem" is the collective-action problem—how do we organize society to work effectively together to solve problems and make our lives better? All of our great systemic challenges are also collective-action challenges. Markets certainly have a role to play, but we need to understand and employ the full ecology of institutions required to effectively solve our problems: markets, states, and civil society. As several of the panelists note, framing these institutions as in opposition to each other has not only been misguided, but has also been very damaging—it has reduced our capacity to solve our big collective-action problems.

Margaret Thatcher was wrong; there is such a thing as "society." It is an emergent property of all of us working together, harnessing our cooperative, prosocial instincts to solve the shared problems that we all face. The economy, as Karl Polyani observed, is fully embedded in wider society, not something separate from it. Thus, when things that are good for society—stopping ecological collapse, reducing injustices that create conflict and division, or preventing technology from destroying democracy—are portrayed as somehow "bad for the economy" it just exposes the absurdity of seeing the economy as a mechanistic system of atomistic individuals somehow separate from the rest of society. Many economists understand this and are working hard to bring "society" back into economics (or, more appropriately, economics back into society). There is new energy to bring a richer understanding of institutions, the evolution of cultural norms, social psychology, political economy, and issues like race, gender, and power back into the mainstream of economics.

I say "back" because earlier economic traditions grappled seriously with these issues. But now is an opportune time for a major update of our thinking because a number of fields have made exciting progress on different pieces of this interdisciplinary puzzle. The Santa Fe Institute is an ideal community to help put those puzzle pieces together. The stakes are high—if we are to overcome our systemic challenges then a systems perspective on the economy and society is essential.

Now, if you bought my message yesterday that prosperity comes from human cooperation to solve human problems, and progress is the improvement in those solutions over time, then I would argue, that changes the view on both the role of markets and states. In that view, the real genius of markets is not necessarily efficiency and allocation, but, rather, creating an evolutionary competition for solving human problems. The primary job of business is to solve human problems, and the primary job of the state, then, is to create the conditions for large-scale cooperation to allow that process to happen. That involves things like promoting inclusion. You can't have large-scale cooperation if you're systematically excluding large groups of people. You have to have fair social contracts so that the basis of cooperation is fair and sustainable. You have to have institutions that build trust, like the rule of law, money, regulation. And also you need the state to help solve collective-action problems to kind of grease the wheels of cooperation.

Finally, you could add that, in a democratic system, people have a right—even a duty—to define the fitness function that the market system is evolving toward. So, issues like, you know, social equity, climate change, other things are valid social concerns that should be reflected in that market fitness function.

I would argue that we should worry more about creating a healthy ecology that promotes real, tangible improvements in people's lives and less worrying about kind of theoretical trade-offs and efficiency in the system.

So, rather than thinking of them as competitors, think of them as a joint system, as Scott Page mentioned in his talk. He's done some work with Jenna Bednar on this. This ecology has lots of different forms of organization, not just markets and states, but a broader zoo of forms of organization. And we need to consider that system in the round. And I think Santa Fe and the complex-systems view could really help us get much deeper insights into that.

MATTHEW O. JACKSON It's great to follow Eric on this. First of all, I think it's useful for us to step away from labels like "economists" and "physicists," and sort of bashing caricatures of economics and, instead, really focus on what it is that we think is broken and where we can make a difference and how we can improve the world. I like John Miller's framing of things this morning in terms of knots and ropes. We look at the system and we're trying to understand the system, and we can look at individual knots and try to understand how behavioral problems are there. Behavioral economics has done a lot to expand economics from a system where we had rational actors and systems that were frictionless to situations where there are problems in terms of rationality and computation.

> . . . the real genius of markets is not necessarily efficiency and allocation, but, rather, creating an evolutionary competition for solving human problems.

I think the other aspect of it, and part of this panel in terms of architecture, is that we care about how the structure matters in terms of who's acting with whom, how are they interacting, how does that matter? And I think that Cosma put it well when he was talking about Becker's description of how constraints in markets make a difference.

Actually, sometimes actors have really very little to do with things. So, agent-based modeling can be very important in understanding systems where actors and individual actors make a difference. But there's a lot that's governed by structure itself, if we want to understand financial markets and where the failings are. These are fairly large structures, and we can see a lot that's structurally going on without actually getting into too much detail on the individual level. So, I think that there's room for this kind of level of thinking.

Secondly, I'll sort of echo this idea of the first welfare theorem, which really comes out of Marshall and then gets elaborated on by Pigou, and eventually was really put into place by Lerner in the Lange–Lerner debates with Hayek and von Mises about whether socialist economies can function. And it's interesting that the first welfare theorem about how markets are efficient actually came from somebody trying to push the socialist perspective. It was Lerner who really formalized it. But the idea there is, you know, markets efficiently allocate things. They get prices right. They move goods around. And what's wrong with that? When does it go wrong?

It allows Scott to buy his socks. It allows us to get these chairs that we're sitting on. It does a lot of really wonderful allocation. But there are two things that it doesn't do. One is, it doesn't talk about who gets what. It doesn't talk about the distribution of profits and rents that occur in the economy. The second is, it doesn't talk about externalities. So, it applies in worlds where whether I'm sitting on this chair doesn't affect whether somebody else is sitting on their chair. Or what color socks I buy doesn't cause a shortage in the sock market.

When you look at financial markets, those are different. Those markets don't function in the classical way where you get efficient allocations and efficient movements. And why is that true? Counterparty risk. This term is something that's real.

So, if I take a position and I have debts to you, and I take a wildly risky portfolio, and I end up going bankrupt, that causes problems for your business. And that's fundamentally different from the kinds of markets that operate well. Understanding the architecture then is important, because if I default, we have to know who it is going to hurt, right?

We want to understand how these markets work. From a complex-systems perspective, the easy kind of answer is to say, this looks like contagion, right? We know about how contagion works—we see flus, we see things like that. And we can talk about movements of capital around just like disease and so forth.

I think when we really get into these, this is where doing our homework matters. Trying to understand things in detail matters. I think of an analogy—if I was a soccer coach and I think about soccer, and I'd say, "Okay, look, there are people. They're moving a ball around, they make passes, they try and get it in the net." Okay, so I understand soccer. And now I want to go and I see basketball. I see people moving a ball around, and then eventually they're trying to get it in the net. And I say, "Well since I know about soccer, I can take that knowledge and deal with basketball."

Obviously you know when we look at these kinds of simple contagious systems, and we get to financial economics, there are subtleties and specific things that go on there that make it fundamentally different from standard flu contagions.

In a financial market, if you actually densify the connections, you're diversifying. At the same time, you're allowing things to spread more easily. You're also making people less likely to spread.

And so there are fundamental differences in the way that those markets work from the way that other markets work. That means we have to adapt our systems to that. I think understanding these architectures, understanding enough of the detail to know how we want to apply these things, is really fundamentally important.

ALLISON STANGER Thanks. Well, I'm Allison Stanger, and I really like what Brian said to start off our panel. I thought I'd begin there. It's a great question to ask. Can we do a better job of anticipating who is going to game the system? When you think about how you would deal with that problem of all systems potentially being gamed, to preempt that, you really do need government. You can't get away from government because self-regulation won't work for a variety of reasons.

So the strategic counter system is smart government. Our problem, of course, is we have anything but smart government right now. That's a whole separate matter. So, I thought I would come at this question from a slightly different angle and maybe start with a joke that I heard as a graduate student at Harvard.

I'm a fallen economist, by the way. I was very influenced by Larry Summers's 1984 NBER Paper on the scientific illusion in macroeconomics.[2] I don't know if any of you remember that paper. That really influenced my thinking, but the joke at the time was what happens to a mediocre mathematician? They become a brilliant economist. What happens to a mediocre economist? They become a brilliant political scientist.

[LAUGHTER]

I'm an example. I started in math, went to economics, then went to political science. So, I'm fallen, in some sense. I want to suggest, however, that if we get more voices in the room, we can think more creatively

2 Later published as Summers, L. H. 1991. "The Scientific Illusion in Empirical Macroeconomics." *The Scandinavian Journal of Economics* 93 (2): 129–148.

about economics. And we're seeing a little bit of it here today and it's indeed one of the foundations of SFI.

I want to make three points here. And they're really focused on complexifying the agent him or herself. If we're talking about the idea that's done the most harm, many people are focused on the rational-actor model, *Homo economicus*. I would add that neoclassical economics, like many Enlightenment projects, was unaware of its own blind spots. It abstracted away key aspects of the human condition. Namely here I mean the human desire to connect and to belong to something. So, *Homo economicus* therefore needs to be supplemented or replaced with what I would call—and this comes out of work we did with Anne-Marie Slaughter and Hilary Cottam, whom Scott mentioned—*Sapiens integra*. Basically, *Homo economicus* seeks to maximize individual economic utility, whereas *Sapiens integra* seeks to grow capability: their own, and those of their networks. They care about both individual and collective flourishing.

So, what does that mean for the ecology of any economic architecture? What that means is that it must include both types of agency and incentives, because, at different points in time, some will prioritize the pursuit of profit and others the pursuit of capability, both for themselves and their communities.

. . . appreciation of diversity's potential role in revitalizing the rule of law on which free market capitalism depends should be of critical importance.

We can talk about what four animating capabilities would be. We mentioned learning or education, health, relationships, and community. But from this you get an idea of a moral economy, which is one that does not render invisible the possibility of *Sapiens integra*, because this might sound . . . I'm the one female on the panel . . . and this might sound like some kind of gendered thing, but I don't think it is. I think all humans seek both these things, just in varying degrees. You know, some who are in the ten percent that was mentioned earlier become economists, but that's another matter. But this isn't gendered.

Final point to ponder: Does *Sapiens integra* of necessity become *Homo economicus* when empowered? Because this is something really interesting to think about. Neoclassical economic models assumed empowered rather than exploited or dominated agents, and that leads to a whole variety of unintended consequences and ramifications.

The economy, I think, looks very different depending on where you sit in a socioeconomic hierarchy premised on wealth. And the character of that hierarchy obviously raises inevitable political questions.

So, I think appreciation of diversity's potential role in revitalizing the rule of law on which free market capitalism depends should be of critical importance. Some of Scott Page's work on this is very interesting. If we're looking for better outcomes, if we're looking for a healthier ecology, this might be one way to do it.

Just to conclude, I wanted to endorse an earlier idea that was first suggested in an audience comment on an earlier panel, and then I think Josh liked it as well. Eric did, too. This is the idea of an open-source public library of cognitively plausible agents. I think this could be a great bridge to both the policy world and more humanistic areas of inquiry, since both agents and economies have histories. This means that humans might be conceived of as emergent phenomena, as well as economies. This is a kind of neat way to get back to Geoffrey's point about bringing velocity and time back into our models and our ways of thinking. How's that?

[ALLISON POINTS TO TIMER AND IT GOES OFF; LAUGHTER]

W. TRACY One of the common themes that's emerged, not just in this panel, but in many of the discussions we've had, is the sense that the current system—the current economic architecture—is unsustainable. But, despite economics' moniker as the dismal science, there seems to have been an almost groundswell of optimism for how we can prove it and what could come next. I'd like to push all of you to consider, as we go into this transition: What are some of the potential pitfalls? What are some of the architectures we could end up with that would actually be as unsustainable as what we have now, or in other ways worse than what we have now? What are you afraid of as we go into a sort of transition of how we organize or architect our economy?

W. B. ARTHUR I have to tell you, to go back to what Geoffrey West said, I'm much more worried about the environment and climate change and possibly also about the future of AI than I am about the economy. But let me give you one thing. This is not continuing the theme that I was on earlier. Just very quickly.

I think the coming of artificial intelligence and the taking of a lot of economic functions and putting them into a digital, rather invisible economy that executes largely without human beings is on its way. Eric mentioned John Keynes's article in 1930, "Economics Prospects for Our Grandchildren."

Keynes first saw that we'd have a lot of technological unemployment 'round about now, or slightly in the future, a hundred years. And that we might be in a situation—I think this is coming—where, in principle, we have enough goodies or wealth to distribute among everybody. We're well off as an economy whether it's the US or Japan or Germany or anywhere else.

How do we get by in an era where distribution of wealth becomes the dominant problem?

But the problem is that people don't have access to that if jobs start to disappear seriously, and I think they possibly will. Then having all that wealth is not going to be that useful if people can't have access to it.

So, to answer your question, I think we're moving from a situation in the economy where the primary problem is production. Can we get more stuff produced, more goods, more services, better ones? That's always going to be a challenge. We're moving from that into how do we manage to make sure everybody has access to whatever's there. We're moving into a distributive era. And that means questions of efficiency are less important for the questions of who gets what.

And that's causing huge political disruption. Not just in this country, but all across Europe. I can feel it and see it to some degree in China as well. And in Asia. And it will affect all advanced economies.

How do we get by in an era where distribution of wealth becomes the dominant problem? We haven't solved it.

E. D. BEINHOCKER Like Brian and many others, I share huge concerns about the twin stressors of climate change and huge technology disruption coming. But I have a meta-concern overlaying that, which is our capacity to deal with those challenges—in short, our ability to solve collective-action problems and work together to figure out how we address those.

Many of you know Robert Putnam's work—famous political scientist who's written a lot about social capital. He's got a new book coming out. I saw a preview of it, and he had a lot of data showing a huge breakdown in our social connections. This is building off of his famous *Bowling Alone* work, and a rise in this kind of hyperindividualism, in attitudes, particularly across Western societies, which, he argues—I think rightly—has reduced our capacity to work together.

There are multiple explanations for this, but the kind of economic ideology that's been dominant has emphasized that pursuing self-interest is good. You asked yesterday about the most dangerous ideas in economics—I would add to the list "methodological individualism," which is this idea that we're all just individuals working in isolation, only interacting transactionally. This led to Margaret Thatcher's famous quote, "There's no such thing as society." If you stop believing there's such a thing as society, you stop believing in the ability of society to act together to handle a challenge like climate change or technology.

At the root of this is a confusion between collectivism, which the neoliberals critiqued, quite rightly, versus solving collective-action problems. One of the big successes in the Enlightenment project was an emphasis on individual freedom, individual rights and so on, a huge jump forward for humankind. But then, taking that to an extreme, we lose sight of our ability to work together to solve collective-action problems. So, we have to kind of rediscover ways to create an ideology and system of thought that allows us to tackle these huge existential problems together.

M. O. JACKSON One thing I touched upon yesterday was that economic systems interact with social systems, so if we don't understand the social fabric, then we miss a lot of what's happening in the economics. And that has to do with access to education, access to jobs, a lot of things that I've spent a lot of time researching.

I think the other answer gets back to Geoffrey's question about why people aren't leading the charge in climate change; there was a letter signed by economists, calling for higher carbon taxes. I signed it. There are debates about exactly how high those carbon taxes need to be and what level they should be at. I don't think there's any disagreement among a lot of economists about what needs to be done in order to get reductions in carbon emissions and to move things in the right direction. I think what's missing is that layered on top of this is a political system, and we have markets interacting with the politics. Unless the politics are willing to make long-term, very costly decisions that will impose taxes to actually bring things into line, nothing's going to happen.

And so, somehow, saying, "Okay, we just have to get the economics right"—that's not it. We really need to understand this as part of a big system. And the big system involves people having vested interests in politics, politics not necessarily representing the majority view in a world of "we have a crisis and we need to solve it," and it's a long-term crisis.

I think we don't have the models for these kinds of problems yet. We don't really understand. If we want to get something improved on that level, we have to have some better understanding of the interaction between politics and economics and social structures.

A. STANGER Excellent. I couldn't agree with you more. I'm an outlier in another way. I spend a lot of my time trying to intervene in the political system to produce some results. Because what I see in Washington is that everybody there could be fired if they don't tow the party line. The great thing about tenured academics and why they should be involved more in politics is that nobody can fire you for saying what's true.

I feel an obligation to do that in the work I do. Once you say it, it's in the congressional record, and that frees up a space for things to happen. I'm not trying to minimize the problems we face, but that's something we can consider.

DENSE FINANCIAL NETWORKS: RISK & REWARD

Matthew O. Jackson

It's hard to find a better example of an important "economic architecture" than a financial network. These networks move capital from those who have some to invest to those who have needs and opportunities. Without financial networks of intermediaries, our economies would not function. These are now global networks, with well more than $100 trillion of investment coming from foreign sources—more than a third of the world's total investment.

Global financial networks have played an important role in enabling the enormous growth and reduction in poverty that the world has experienced since the Second World War. At the same time, these large structures expose us to huge systemic risks. As we saw in the 2008 financial crisis, bankruptcies and defaults in one part of the world can have global and long-lasting consequences.

They pose problems because financial networks are full of externalities. For instance, one bank's poor investment decisions can lead it to default, which then imposes costs on its counterparties. The external ramifications of the bank's poor decisions affect someone far from the bank, in ways that cannot be insured against. Some critical niceties of Adam Smith's "invisible hand" are missing from financial markets, which makes understanding their operation so important.

Financial networks are complex systems that sit well within the Santa Fe Institute's wheelhouse. Small changes in one part of a network have ripples and far-reaching, cascading effects. There are, however, important twists that make these systems different from ordinary contagion networks. Here is where the "economics" enters. As one example, making a financial network denser and more connected can actually make it safer, quite the opposite of what happens with a disease network, for which increasing interactions leads to greater contagion. This is because as a bank spreads its business among more counterparties it becomes less susceptible to the problems of any one of them. Combining economic insights with modeling techniques familiar to SFI is essential to this area of study.

Global financial networks have played an important role in enabling the enormous growth and reduction in poverty that the world has experienced since the Second World War. At the same time, these large structures expose us to huge systemic risks.

An important point that came up in our panel discussion is that such systems cannot be viewed in isolation. This echoes the main point of my earlier presentation in this symposium about interacting complex networks. Here, one cannot understand the functioning of financial networks without understanding the politics that drives their regulation. Historically, we have seen political cycles that have reacted to crises and driven regulation quickly upward, only to then erode over time until the next crisis.

The interactions between economic and other systems underlie many of the world's most pressing problems. For example, inequality has received enormous attention in recent years. Much of the discussion has been on redistribution, and to some extent on capitalism. This misses the point that one cannot understand inequality in economic outcomes without understanding social structures. The homophily and segregation that we see in social networks play central roles in driving inequality in education and opportunities, which in turn lead to eventual wealth and income inequality. Redistribution and rewiring economic markets deal largely with the symptoms, but ignore deeper roots of the disease.

One of the most significant challenges that I see in dealing with these sorts of large-scale economic problems—from financial crises, to pandemics, to inequality, to climate change—is the interaction with political systems. We have enough knowledge about many of these interacting complex systems to suggest policies that would dramatically improve the world. However, standing in the way are a wide variety of vested interests. When coupled with polarized world views and paralyzed or dysfunctional political systems, it is hard to get the necessary consensus to act with the urgency and scale needed to address our planet's most pressing problems.

Further Reading

Jackson, M. O., and A. Pernoud (2020) "Systemic Risk in Financial Networks: A Survey," SSRN preprint 3651864: http://ssrn.com/abstract=3651864.

I'm afraid of the same sorts of things that you are. Not to sound like Roosevelt, but I'm also afraid of fear itself, because it's contagious. And if we think it's hopeless, we're going nowhere. So I refuse to give in to that. I'm also very afraid of the growing social inequality we see that I believe the digital economy, as Brian mentions, is simply going to reinforce and exacerbate. Thinking about how we're going to address that is an enormous challenge.

. . . we can interrogate our own blind spots. We're kind of in a call-out world where there seems to be an assumption that only certain groups of people have blind spots. They're human.

Here I have three suggestions. They're very practical. You can apply them at your company or wherever else you may be working.

One, you can question inequality of outcomes whenever you see it. It doesn't mean there's some sort of bias or discrimination there, not at all. But you can at least step back and say, "Hmm, why do all these people look the same?" or, "Do they have the same degree?" and see if maybe you're missing some excellence in that system you haven't seen previously.

Second, you could do the same thing with any table where you're discussing these issues. Be sure that there's diversity around the table. I mean this not only in terms of different disciplines, hugely important, but also in terms of civil society, government, and business. The biggest problem I had with the Partnership for AI was that it entirely left government out of the discussion, and that's not going to produce optimal solutions.

Finally, as I said earlier, we can interrogate our own blind spots. We're kind of in a call-out world where there seems to be an assumption that only certain groups of people have blind spots. They're human. They're so human. I just had one of mine revealed to me the other day with a panel at New America for my book where we were putting together a list of potential discussants. Suggestions flew. "Oh, this person's really smart. This person's really smart. This person's really smart." And then I realized: "Oh, wait, they're all white men." Somebody interjected, "Can't we do better than this? Isn't there somebody of color that might have something important to say?" And you know what? All I had to do was stop to think about it. And I realized, "Oh my God. I have the perfect person." She wound up stealing the show on the panel. So, I just offered myself up as an example of a person with blind spots as well.

W. TRACY I want to open this up to the floor and make this an interactive conversation. Do we have any questions for the panelists?

BARBARA GROSZ[3] I'll start with a disclaimer, which is I do research in AI. And, Brian, I'm much more worried about the economic models than I am about AI, and I'm worried about the connection between them that other panelists have brought out. I want to start first by recommending Mary Gray and Sid Suri's book, *Ghost Work*. By "ghost work," they mean all of the people who do "gig" jobs on the internet. You would think they worked solely as individuals, but actually they form communities and advise each other. These jobs are terrible. They don't let you work when you want to work, but, rather, they require workers to pay constant attention to what's going on so they can get the good jobs.

We need to think about the jobs that technology is enabling and disenabling so that corporations can save money—unlike before the 1950s—without thinking about the well-being either of their employees or of those of us who are customers. Companies now count on free labor from us to deal with their bad systems.

It's wonderful what AI can do, but it's nothing like you might think from reading all of the brochures and advertisements. There may be AI systems that will take over jobs, but they won't be doing them as well as the people who are doing them if they were designed just to take over the jobs, as opposed to complement the people who are doing those jobs. We see this in medical care now with electronic health record systems that help a lot with the billing department, but not with healthcare delivery.

3 Former SFI Faculty and Science Steering Committee member; Higgins Research Professor of Natural Science, Harvard University.

So, I think we have to fix the economic system and the economic incentives in ways some of the other people were speaking about on the panel. And that's going to take the political input that Allison was talking about. So, I think you should deal with your anxiety about AI by fixing the rest of it.

[LAUGHTER]

W. B. ARTHUR I wouldn't say it's an anxiety about AI. I work at Xerox Parc. I am concerned that we've had, since the early 1990s, in fact—due to telecommunications rather than computation, we were able to off-shore jobs to China, to Ireland even, to Mexico, and that led to much cheaper products. You could keep track of supply chains in real time, etc. And it looks like possibly some of that's ending. At least, there's a political backlash to that, but now we have a new country we're off-shoring things to, and that's the virtual economy.

I don't feel I've an anxiety about AI; I've an anxiety about jobs disappearing. If it's true that an awful lot of trucking in the next ten or fifteen years will be done autonomously, a friend of mine in Silicon Valley has calculated that three million jobs are at stake. Not just truck drivers, but all the pit stops and the facilities, the gas stations or whatever that depend on these—a lot of that will be automated as well.

So I think that AI is going to bring us brilliant insights maybe into medical diagnosis, possibly allow us to learn individually. The benefits are going to be enormous. My anxiety is that we're not thinking enough about it. In general, science thought a lot when recombinant DNA came along in the 1970s, and got together and came up with protocols. I don't see we're doing that for the disappearance of jobs via automation, AI, computation and so on. And I support all that, but I think we need to have second thoughts.

AUDIENCE MEMBER There are a couple things that are disrupting the architecture of the economy in tremendous way. One of them is the shrinking or the disappearance of the middle class, which creates the continuum for the flow of wealth across different classes of society. The other one is the increasing number of unemployable people thanks to advances in productivity. How do you see these two forces affecting the architecture of the economy and what can we do about it?

E. D. BEINHOCKER You put your finger on two big issues. You know, the shrinking of the middle class has been quite dramatic. We've only quite recently gotten the right kind of data where we can really see this over time and quickly how far back it goes. It goes back well before even the kind of globalization and outsourcing stories of the '90s or the tech stories. It goes right back to the 1970s, so it's been going on for quite a while. We also know that middle classes historically have been hugely important for economic growth and innovation. In fact, an economic historian has written a series of books about why this was critical to the Industrial Revolution, the development of the middle class.

It's the middle class that consumes stuff, that has the jobs. Middle classes are also essential to democratic polities as well. So the shrinking has had both an economic effect and a political effect. I should be careful with a real political scientist on the panel, but some of the political-science research on the recent political trends has pointed to fears of downward social mobility, loss of status, loss of dignity in work, things like that as being crucial in creating a lot of the anger that's out there. This isn't one side of the political eye or the other; it's really across both.

We've done some recent comparative work across countries about the shrinking of the middle class to try and get a better handle on why, and it really does look like it has a lot to do with policy choices. That all the developed countries are facing the same pressures from technology change and from globalization, yet some have held onto their middle classes much better than others. So, there is an economic kind of architecture and policy choice to support and promote a healthy middle class.

Looking forward with the technology-change issues, you can see a huge acceleration in that hollowing out coming. We recently did some work looking at the skill transitions—which jobs are likely to get wiped out by technology?—and then the skills those people have—where could they go?

Part of the argument is, technology destroys jobs, and it creates jobs, and over history in the wash, it all kind of works out. But there are two things that appear to be different this time. One is, the speed of its change used to be at a generational clock speed, so maybe you'd have a problem, but your kids would be okay. Now it's happening within one lifespan.

The second is that it's wiping out a set of skill areas that have nowhere else to go. We haven't properly dealt with the last big round of disruption, you know, all the left-behind places from industrialization, whether it's in the industrial Midwest or the suburbs of France, or the coal-mining districts of Wales. We still haven't dealt with that problem, and we're now creating a whole new set of those issues on a much, much bigger scale.

> We don't have to be addicted to GDP growth in order to create prosperity.

OLE PETERS I have a comment. I wanted to take a step back and just share what I'm getting out of this meeting. I noticed two really big important themes that have, thank God, come up. One is climate change and its relation to growth. The other is inequality and probably its relation to growth. And I just wanted to say it sort of makes me think, maybe I gave the wrong talk yesterday. Please do not underestimate the power of foundational work, of going back, really to basics, and starting over again.

In what we've done, there is no equilibrium assumption. It would have never ever occurred to us to put equilibrium as an assumption into our models. You've seen the trajectory yesterday. We've had a few man-years to explore what happens when you start in this completely different way. Two big things that come out of this are the problem of climate change and the problem of inequality.

GDP is in essence an ensemble average. It doesn't tell you what happens to the individual. We don't have to be addicted to GDP growth in order to create prosperity. This is just not true; that link doesn't exist. And this comes immediately out of what we're doing. So this is the climate-change thing. Of course, if GDP is not connected to well-being and you could see how things disperse and how a wealth distribution may be completely unstable, you end up with a completely different mindset. These problems that have come out luckily over these days are kind of the first things that you notice when you start from this perspective.

JOHN CHISHOLM[4] Eric, I find the idea of markets and states being components of one large ecosystem very appealing, with the most interesting part being how the state evolves. I see common law as co-evolving with society. With common laws, many decision points and trials distributed across time and geography. But, over the last century, the US has shifted from a primarily common-law country to one primarily of statutory law, which evolves much less readily than common law.

And when I look at the services states and the federal government provide—K–12 education, the US Postal Service, the Department of Motor Vehicles, I could go on and on—these seem to evolve least of all. Of course, they're also all subject to public-choice concerns, in which factors other than public welfare intrude on decision-making.

I noticed we were dismissive of how bad the current administration is. But the fact is, we cannot rely in general on having a good administration. So, how do we ensure that the state evolves in a good way?

W. B. ARTHUR Political question.

[LAUGHTER]

E. D. BEINHOCKER Yeah, I'll punt that one over to the political scientist.

. . . because of things very specific to the American situation, our ideology, we degraded the capacity of particularly our civil service over decades, and we're kind of now reaping the harvest of that.

A. STANGER No, that's the $64 million question. I think what's really interesting in this current crisis is the complete misunderstanding that people even have about what the state is. You know, there's this idea that the deep state is part of it. I've been going around the country telling people that, no, the state seems partisan right here at this moment in

4 SFI Trustee and CEO, John Chisholm Ventures.

time, but their behaviors are a radical departure from how they've always previously behaved, with some exceptions.

They see themselves as serving their country and speaking unvarnished truth to power. For them to be leaking things in this way is precisely because they're very alarmed about this particular White House and what he's doing to the very idea of public service to which they devoted their whole lives. Because, whatever you want to say about the civil service, they're not in it for the money. They could all be doing something else and making more money. That whole idea of serving the public has really been damaged by the present administration.

Now, again, this is not a partisan statement; it's a statement about the system itself. I think for us to get to a better place, we have to educate Americans on that fact as a necessary condition for them understanding certain other things about how American constitutional democracy is supposed to operate, even though it often falls short. Because I still think if you're talking about the best operating system for any regime, you can't beat the American Constitution. We have to make people realize that.

E. D. BEINHOCKER My sister actually works in the current government and daily tells stories of civil servants really trying to do things for the American people. So that's often underappreciated because the news media narrative is always very, very negative about the capacity of government to actually do things.

When you look around the world at the capacities of different governments, you see no government is perfect, like no large corporation is perfect, by the way, either. But some countries have very effective civil services, and governments are able to take effective action. But, because of things very specific to the American situation, our ideology, we degraded the capacity of particularly our civil service over decades, and we're kind of now reaping the harvest of that.

Frank Fukuyama has written some very good things about this that many of you may have read. But then the hard question is, how do you build back from that degraded state? To do that, you need trust in the system, because people, if they see a poorly performing system, they're not going to trust it. So, you have this chicken and the egg; they're not going to invest in it or support taxes and other things going into it.

My own view is, we have to look at where trust is living today. And where there is still trust today is much more at the local level, you know? Surveys show very high levels of trust in local government, state government. One of the beauties of the American Constitutional system is this kind of devolved power structure that we don't take advantage of enough. And so I would argue we need to sort of build back state capacity from the local back up to the broader collective. It's pretty urgent to do.

M. O. JACKSON I have one thing to add. I think that part of it is ignorance. It's very easy to be ignorant of all the things that the state does. There's actually a really wonderful website that's run by some computer scientists at Stanford.[5] You can go in, and what it does is it gives you actual budgets. And then you can go in and put in your own budget.

How would you design your budget for your national government, state government, local government, and so forth? You get some impression of actually how much spending goes into each part, and what all these pieces are and how much have to go to roads, and how much have to go to K–12 education, and so forth.

Once you see those numbers, you begin to actually see the trade-offs. I think most people don't have an idea of the scope or the size of the state. Just having that idea then gives us a little more trust, or a little more understanding of what it does for us, and instead of just attacking it, understand all the benefits it provides for us.

A. STANGER Just so you see how unusual the situation is, I was supposed to speak at the State Department last Monday, subject to the approval of the Secretary of State. I was not approved. That is not the United States of America.

AUDIENCE MEMBER Hello. I lead a design and construction company. Are any of you guys in this room actually modeling the revolution? I mean that seriously. Yesterday, Mónica gave a talk about the increase of demonstrations around the world, different types of protests, anti-state movements, things like that. And it seems to me it's kind of like the

5 https://pbstanford.org

elephant in the room. I mean, the recent article in *The Times* says that we have underestimated the rate of climate change, and there may not actually be anything left to model by 2050. So, I would be interested to know if anybody in the room is modeling the revolution. What are your assumptions? And what sort of conditions and when should I start buying my guns, so to speak?

[LAUGHTER]

E. D. BEINHOCKER Sorry, say more what you mean by revolution.

AUDIENCE MEMBER Social collapse—1776 or 1789 or Germany inter-war or something like that. You know, these revolutions that will lead to the ultimate destruction of all the capital stock.

A. STANGER Except the Bitcoin.

[LAUGHTER]

E. D. BEINHOCKER You know, I don't know—maybe Josh. Have you been modeling that?

JOSHUA EPSTEIN I have a paper in *PNAS* called "Modeling Civil Violence: An Agent-Based Computational Approach."[6] It models social collapse through the accumulation of grievances and illegitimacy and economic hardship. It models both ethnic war and rebellion against central authority.

AUDIENCE MEMBER What are you predicting?

[LAUGHTER]

J. EPSTEIN I'm trying to understand the mechanics of social collapse and civil violence and questions of legitimacy and economic hardship. One interesting thing is that sharp declines, shocks, in legitimacy are much more likely to produce revolutions than much larger slow incremental reductions in legitimacy, for reasons that are quite interesting. But this is why many rebellions begin with a single incident. The assassination of the Archduke Franz Ferdinand, fraudulent election results, and so forth. It's because a lot of people go into the streets at once.

6 Epstein, J. M. 2002. "Modeling Civil Violence: An Agent-Based Computational Approach." *Proceedings of the National Academy of Sciences* 99 (Suppl. 3): 7243–7250.

Once a lot of people are in the streets, it's much less risky to join the mob. To be the first actor, you have to be really ticked. But to be the thousand and first, you don't have to be. So the trick is to get the thing, the shock, that will move a lot of people at the beginning to amplify.

W. B. ARTHUR A word about revolutions: there was a major revolution right across Eastern Europe and into the Soviet Union, 1989 through about 1991. At least from the point of view of the West, we don't regard that as something of a collapse. It was maybe a reset to how those regions were run. I could give you my argument in one sentence. I come from Ireland. We have had a revolution there since the Middle Ages just about once every generation or once every fifty years. The United States had its own revolution as well. These were regarded as something for good.

What I think is going on is a profound malaise or an unease about prospects for employment and prospects for doing well. And that unease seems to be everywhere, from Algeria or the Middle East on through to highly developed countries. And that in turn is bringing a dissatisfaction with the status quo, which is bringing these bubblings up of so-called revolution.

What I've noticed here is that there's a major forgetfulness in this country and that revolution is typically regarded as something reprehensible and to be avoided, yet the country is founded on that.

J. EPSTEIN The other thing that's interesting about social media in all this is that, if you look at the Arab Spring or the Hong Kong rebellions or other uprisings, they're leaderless in the sense that there's no identifying Mao Zedong or V. I. Lenin or other such people. I think the capacity for self-organized revolutions is a very important development enabled by social media. I also have a little toy model about the Arab Spring that tries to incorporate that. But I think that social media is ushering in an age of leaderless revolutions.

W. TRACY I'll take one more question before we wrap for the day. Tim?

TIMOTHY KUBARYCH[7] Thank you. Kind of on the same topic, collective-action problems are really hard—but what should be the

7 SFI ACtioN member and Partner, Deputy Director of Research, Harding Loevner LP.

BROADENING ECONOMICS' SCOPE

W. Brian Arthur

This panel surprised me. We were asked to think about economic architectures—the role of markets, networks, structures, and institutions in the economy—and I expected the discussion would be quite technical. Instead, what emerged was a far-ranging collection of issues and concerns that need attention by economics: climate change, social inequality, lack of diversity, the coming of artificial intelligence, loss of jobs, the gaming of economic systems, economic collapses, the unrealistic representation of humans in economics, the flimsy relation of economic thought to political thought.

Could a broader economics deal with these issues of concern, and what sort of economics would that be?

These look like a motley collection of worries, but they do have something in common: To a high degree they have been left out of economic discourse, and this is because they don't fit the standard economic framework, meaning that they don't neatly fit with market allocation theory. At best they become externalities or outside concerns with (as Matthew Jackson said in the panel) politics layered on top. Standard economics is not directly at fault for this. It developed over the last century and a half to analyze problems of allocation in markets using algebra and calculus and based on assumptions of rationality and equilibrium. And indeed it has successfully dealt with how production and consumption work in markets, along with topics such as monopoly, taxation, the workings of financial markets, the role of information, intergenerational transfers, international trade, and other topics. But it has not dealt with problems that do not fit this equilibrium schema—how economic policies can be gamed, what causes economic collapses—and this has set them outside standard economics. I believe this is in no small part why they remain ongoing problems.

Could a broader economics deal with these issues of concern, and what sort of economics would that be? It would have to be an economics that would get away from perfect markets as a benchmark, and would not necessarily count major issues as mere externalities and possible market failures. It would accept that agents differ, circumstances differ, markets differ. It would include realistic agents working realistically within networks and fabrics of social arrangements. It would recognize that incentives are important and that these change endogenously over time, and that institutions and systems also change over time. It would not be based on stasis or equilibrium. It would see agents' beliefs and actions and strategies competing or cooperating in an ecology of other agents' beliefs and actions and strategies. It would in other words see the economy as an evolving, complex system, and in this way it would broaden economics' ability to analyze issues the standard theory does not. It would be complexity economics, or something very like it.

One development that makes this new economics possible is the coming of computation within economics. In the early 1980s we all got computers and this allowed us to break free of the strictures of standard economics. We could of course still use standard mathematics, but now we could also set up computational models to track agents that differed and had behaviors that evolved over time, and track how structures and patterns and outcomes would endogenously form and continually change the economy. This development isn't yet universally accepted in economics, but it is valuable nonetheless. Computational models allow more than agent-based behavioral realism, they allow realistic detail: standard economics typically relates average aggregate quantities (outputs produced, say) to average aggregate quantities (inputs used) and often breaking out such aggregates matters.

I don't think that complexity economics and its ability to use computation will answer all the problems the panel worries about. But above all, complexity widens economics and allows it to look at systems—and issues—evolving over time, affecting each other, bringing structural change, and forming within ongoing nonequilibrium. With these new possibilities I believe economics will broaden the problems it can deal with.

action is also hard. A lot of people in this room who are very thoughtful and very learned on certain topics, it starts to sound like, "I know a lot about ecology; therefore, I should be the one that makes decisions about climate change," whereas I would think that it's really hard to know what you would do about climate change. It's a very complex adaptive system, and it's very challenging to know what to do with the Euro, or know what to do with AI and job loss and distribution of wealth. Those are topics that are really hard and not knowable and unforecastable, that lead to unintended consequences.

I was struck when you said before, it's not what the majority of people want. But the way that people talk is not about adjudicating difference of choices within a set of decisions within a collective-action problem; it's binary, like, "I don't support changing distribution of wealth," or "I *support* changing distribution of wealth." But even within each of those domains, there's very little visibility on how you would create a democratic group of individuals to voice an opinion on one choice or the other. What is your opinion on a better way of getting inputs from people or institutions, getting more information such that the choices can be adjudicated instead of the kind of trivial "should we do something about a big problem"?

. . . there's a major forgetfulness in this country that revolution is typically regarded as something reprehensible and to be avoided, yet the country is founded on that.

E. D. BEINHOCKER Would you like to take a whack at that?

W. B. ARTHUR Yeah. I'm often comforted by going back into history. What interests me is the economic setup in England around 1850. The Industrial Revolution had hit by then. There were railroads, there were textile mills. There was an incredible amount of additional production. And yet people were working in quite awful conditions in Manchester in the mills, working fourteen-hour shifts six or seven days a week sometimes, and being very ill used.

What strikes me about that is that it took thirty to a hundred years for conditions to improve. Typically, it's not governments that step in and say that we have to do something; it's more that there are concerned people. They could be nurses or sometimes peers of the realm, Lord this and Lord that, sometimes lawyers—but concerned people. They get in and they say we need to do something about sending little boys up chimneys as chimney sweeps, etc. We really need to rethink what's going on. Bit by bit, the economy or society adapts. There are better labor laws. Trade unions or labor unions came about, at least, got much more power. So, society does adapt.

My question at the moment is, can we adapt rapidly enough? We will adapt, I think, to climate change. The question is, will we adapt fast enough? Part of that has to do with public opinion, but a lot of it has to do with small changes and just getting things done in the right direction.

. . . you make reforms one step at a time. You sort of march your way forward and you try stuff and not everything works but some stuff does work and you do more of what works and less of what doesn't. That's an evolutionary process.

E. D. BEINHOCKER One big thing that Hayek got right was, in complex social systems predicting the future is a dangerous business. Popper's answer to that challenge was what he called "piecemeal social engineering," so he was very much against Utopianism and kind of big grand plans. But he saw that in the historical records, of course, as to Brian's examples, that you make reforms one step at a time. You sort of march your way forward and you try stuff and not everything works but some stuff does work and you do more of what works and less of what doesn't. That's an evolutionary process. And evolution can grope its way toward whatever goals exist in a complex system. One could argue the three great inventions of the Enlightenment—taking into account the blind spots—were three evolutionary systems: science, which is an evolutionary epistemological system; democracy, which is a kind of evolutionary political system; and markets, which are an evolutionary economic system.

But, as I argued yesterday, in recent decades we've had an ideology that has said the only fitness function that matters is the maximization of self-interest and a narrow definition of profits and so on. So, we've had a system that has been adapting to that and giving us the results that we have today. One of the key ways to handle these big challenges is to reinvigorate these evolutionary processes in our society and get them pointed at solving our big societal challenges.

W. TRACY Unfortunately, we are going to have to stop now, but let's thank all of our panelists again.

[APPLAUSE]

GUEST CHECK
689561

6

FROM THEORY TO APPLICATION

[Panel Discussion]

Moderated by Paul J. Davies,
featuring Katherine Collins, Michael Mauboussin,
Bill Miller & Dario Villani

PAUL J. DAVIES My name is Paul Davies. I write about markets at *The Wall Street Journal*. I have no particular theoretical or economic background at all, so I'm enjoying a very hefty dose of impostor syndrome. I'm here to moderate a panel where we will, hopefully, talk a bit more about some of the more practical applications of a lot of what you guys think about and write about and so on. I've long been interested in this area. I guess it all started in the financial crisis. I was lucky enough to be at the *Financial Times* writing about credit derivatives and securitization and all that weird and wonderful stuff.

The vast majority of people that I talked to at that time about the credit bubble said that nothing's going to stop this. The only thing that's going to end this great, wonderful time is some major external blow. War with Iran was one of the things that people would always talk about. But a handful, really only one or two very, very smart people I knew said, "No, the shock will come from within." You know, that it will be—and people knew that it would be—something to do with liquidity. This was going to be the thing that would end it. That would pull, maybe not in quite such a spectacular way as actually happened, but would sort of knock the train off the rails. And so that whole idea of endogenous shocks first brought me into this area, and through lots of different books, and ultimately coming across Eric's book *The Origin of Wealth* is what opened up this whole magical world to me.

OPPOSITE *Economic Curiosity: Lunch with Warren Buffett*

I also write about insurance and catastrophe modeling and how you work out what's going to happen with earthquakes and storms and diseases and floods and terror attacks and all of that sort of stuff. I've written a lot about banking regulation and that sort of thing as well, which is where, actually, a lot of kind of practical use of more heterogeneous ideas of finance have happened. For example, it is interconnectedness rather than sheer size that decides how much capital a JP Morgan or an HSBC should have. You know, that's a real practical use of something that's not really classical economics at all.

So, anyway, we're going to run this panel in a similar sort of fashion as the others. Each member is going to have five minutes opening, and then we're going to roll along from there. We should have plenty of time for audience questions. I guess I'll just start at this end—everybody can introduce themselves. And the opening question really is, where do you see the biggest gap? Between the theory, between what we think we know and understand about the world through complexity economics and similar things, and the practice, what decision-makers really need to navigate the economy. We'll start with Michael.

MICHAEL MAUBOUSSIN Thank you, Paul. Good afternoon. My name is Michael Mauboussin and I'm at BlueMountain Capital and on the board of trustees at the Santa Fe Institute. I'm going to speak from the vantage point of a fundamental investor, so I'm sort of narrowing the vantage point to some degree. As a fundamental investor, there really are sort of two big tasks to tackle. One is to understand or evaluate the expectations that are priced into an asset. I'll use stocks as a common example, but what expectations for corporate performance are priced in. Second is to evaluate the business to see whether that company will meet or exceed that set of expectations. I'll frame my comments in sort of those two areas and what we can learn to close the theory–practice gap. I want to come back to this idea of market efficiency that came up a fair bit throughout the day.

I always get a chuckle out of the fact that in 2013 the Nobel Prize in economics was given to Gene Fama and Bob Shiller, who are polar opposites of this argument. They were given the prize for empirical analysis of asset prices, but Fama being in the efficient-market camp

INSIGHTS FROM FINANCIAL PROFESSIONALS

Paul J. Davies

This panel of financial professionals aimed to address how ideas from complexity economics can be applied to the business of investment and finance. What are the gaps between theory and practice? How can theory improve practice?

It was a panel of diverse talents and experience, but one of the recurring themes was the lack of diversity of thought, models, and strategies in financial markets, and how that causes instabilities.

Michael Mauboussin referred to Blake LeBaron's work, referencing how diversity of investor views seems to diminish as values climb. But while it's easy to talk about crowded trades and everyone doing the same thing, he said, we have very poor measures of that crowdedness and its ecology.

Dario Villani had some encouraging words for all the scientists in the room when talking about the hugely successful hedge fund Renaissance Technologies. Renaissance, with a record of world-beating returns over decades, has somehow cracked the code of finance as a complex system, perhaps by understanding portfolio construction better than anyone else. Still, after more than three decades, no one else has managed to repeat their success. Compare that with atomic weapons, which were replicated within a decade. The one thing he could say about the firm: "They have hired astronomers; they have hired mathematicians; they have hired physicists; they have even hired theologists. They never even interviewed an economist."

The panel covered the peculiar adaptability of markets and how that might make artificial intelligence or machine learning much more important for finance than traditional models. Models can influence behavior and so (like the title of a book by Donald MacKenzie) they become "An Engine, Not a Camera." In finance, when people start to notice patterns in markets, they trade on them and can create a positive-feedback loop, exaggerating their effect. Bill Miller later reminded us that the practice of arbitrage in markets can do the opposite and create a dampening feedback loop.

But when too many investors pursue the same idea—such as the factor-based investing that is an over-engineered version of ideas from Fama and French—the usefulness of that approach can also diminish and the available returns dry up.

Katherine Collins talked about how the synthesized parts of finance—securities markets, trading, investment funds—have significantly outgrown the important base function of finance, which is supporting and developing companies themselves.

In finance, when people start to notice patterns in markets, they trade on them and can create a positive-feedback loop, exaggerating their effect.

She said much of the finance industry has too narrow a view of its strategy and goals. It often becomes a short-term, almost zero-sum game with many investors focused only on small relative gains won against each other. She talked of the fear of entertaining new ideas and how this can make a system more brittle.

Environmental, Social, and Governance (ESG) concerns are already becoming hampered by such an approach, with an efficient but overly simplistic system of scores. She brought a poetic flourish, referring to Goethe's line that "the corpse is not the creature." You can measure and catalogue every element of every butterfly, but it's not the same as seeing one in flight.

Bill Miller offered some much more practical advice: How he has applied things he learned at SFI in the real world. He mentioned Brian Arthur's work on lock-in and path dependence in the economy and how that made him look at technology and invest early (and cheaply) in companies such as Dell, AOL, Nokia, and Apple. He also talked about an SFI meeting on Innovation Evolution, and how that led him to buy stock in Google when it first sold public shares.

The panel discussed homogeneity of risk management and the ways in which that affects the tides of liquidity into and out of certain trades, assets, or funds. But for Bill Miller, all this makes it easier to outperform on multi-year bases—if you can focus on the important things like long-term behavioral changes or patterns among investors, consumers, or companies.

Further Reading

Markets Enter New Phase—Where Cash Is All That Matters. https://www.wsj.com/articles/markets-enter-new-phasewhere-cash-is-all-that-matters-11584546863

Markets Are Calm, Then Suddenly Go Crazy. Some Investors Think They Know Why. https://www.wsj.com/articles/markets-are-calm-then-suddenly-go-crazy-some-investors-think-they-know-why-11562666400

The Market Forces That Propelled a Massive Rally in Long Bonds. https://www.wsj.com/articles/the-market-forces-that-propelled-a-massive-rally-in-long-bonds-11567426486

New Link Between Stocks, Bonds Shows How Markets Have Changed. https://www.wsj.com/articles/new-link-between-stocks-bonds-shows-how-markets-have-changed-11575470950 and Shiller in the inefficient-market camp. By the way, if you are a fundamental investor, you have to believe both of them are right, just not at the same time. The term efficiency itself, by the way, is interesting because that does come from physics, right? How does energy translate into useful work in markets? We talk about how information translates into prices, but it's the same basic set of concepts. Andrei Shleifer's got a wonderful little book called *Inefficient Markets* and there are three classic ways to get to market efficiency.

The first is the one we've been talking about, which is mean variance efficiency. So, you have these rational agents. They understand their preferences, they understand the future, they all agree on the distributions, and then you get rational prices. I think no one in the world believes this to be true. I don't think any economist believes this to be true. However, it's beautiful. It can stand as a benchmark. I would say that it's prescriptively useful, right? It gives you a recommendation of how to behave that can be helpful. And I'll just do this as a survey: how many people in your retirement or in your portfolio own index funds? Raise your hand, please, if you own an index fund. So, you are prescriptively doing what that theory suggests and, by the way, you're all free riders as well. Right? You're free riding off the active-management investment community.

[LAUGHTER]

The second theory is called the absence of arbitrage. If you give a finance professor a couple of drinks at the bar, this is what they'll say they believe, which is: we don't have ever to be rational; we only needed a subset of people to be rational. These are arbitrageurs. They cruise around markets, they're well-capitalized, they buy cheap, they sell dear, and in their wake they leave efficient prices. And again, that's a first order that's absolutely true. But what we also know in the real world is, episodically, these people fail to show up. Either they don't have any access to capital for whatever reason or the cost for them to effect their job is too high.

The third way to get to market efficiency is—the pedestrian phrase would be the wisdom of crowds or complex adaptive systems. And, when conditions are correct, it actually leads to the same equilibrium solutions as the rational model. I would say, under what conditions are markets efficient in this complex-systems approach? There are three. One is sort of heterogeneous agents. By the way, that's why all this behavioral bias stuff doesn't matter all that much. You can have all these little mistakes, but it gets washed out pretty effectively. Second is an aggregation mechanism, some way to bring that information into one place. In double-auction markets—we talked about the ZI traders before. Double-auction markets are extraordinarily good at that. Third is incentives, which is people that do well get more; people that do badly lose, right?

So, we know that markets are hard to beat. We also know that markets go haywire periodically. The question is, why do they go haywire? And the answer is, one of these conditions gets violated. By far the most likely to be violated is diversity or heterogeneity. We're social animals and instead of our believing different things, we correlate our beliefs. We all come to the same view of something.

I'll just mention Blake LeBaron, whose work has been very useful for me. He built a model a number of years ago. There might be new iterations, but he built his agents and taught them finance theory, and then he gave them decision rules of how to trade. He was able to replicate empirical features in the market.

What's interesting in one of his panels is, he showed the asset price on the upper panel and the bottom panel was a measure of diversity of agent, how many rules that were being used. It was fascinating—if you follow just one of the stretches, the asset price is going up while the diversity in the system is going down. Now what's fascinating is people are adopting the same decision rules, and that's important. The first thing to remember is when you initially do this, it reinforces that things are good, right? But we have a panel that shows its diversity.

There are two things about diversity I want to mention. They're really important. The first is that it's a nonlinear function. That's the main thing I want to emphasize. To me the practical thing is to say, in markets we talk a lot about crowded trades; everyone's doing the same thing. I think we have very poor measures of that, so poor measures of what the ecology is. And that would be something that would be cool.

The second thing is that I think there are a lot of interesting things we can do to understand how businesses evolve and, in particular, the role of innovation. Melanie brought up this idea of exploit versus explore, and as investors and as companies we've done an insufficient job to understand some of the dynamics.

KATHERINE COLLINS Hi, everyone. I'm Katherine Collins. I'm the head of sustainable investing at Putnam Investments and also on the board of SFI. I'm also a fundamental investor. Personally, I think the view of investing with an eye toward sustainability is a broader lens, but that is not a universally held opinion. Just to give you a sense of where I'm coming from, my first proposal was to call this the head of regenerative investing, but that was like a step too far for a lot of folks. So, we'll talk about focusing on sustainability as it links to fundamentals, and then the topics of today kind of all come together.

I think there are many different gaps we could explore, but some of the biggest for me sit under the umbrella of where we are here in finance. If I think about my experience over the last thirty years, we've got more and more and more data in terms of volume of data. If you told me all that data was coming, I would have hoped we would be asking much more interesting questions than we're asking right now. In my opinion, we've got this surplus of one particular type of resource and it's actually narrowed the conversation instead of broadening it, which is kind of a bummer, but the potential is still there to widen it and to really be part of this kind of great reconnection.

One theme that kept coming up to me through the discussions this weekend is that all the silos we spent the last century or so creating are gradually starting to come back together. There's tremendous potential in that. There's also a lot of uncertainty and a lot of fear that comes with it.

So, three things that came up in the conversations over the last couple of days that are relevant to my practice and areas where I think we have some great potential to narrow those gaps: One is where we're starting from in finance; the concept of blinders came up in a lot of the different discussions. I think we've got triple blinders when it comes to a lot of the standard tools for investors. One is that the denominator is almost exclusively currency, so it's just finance in–finance out, dollar-based tools. The second is they almost all still relate back to this equilibrium-type model at the core, if you scratch the surface a little bit.

Not at a hundred percent, but pretty close, and mostly they're pretty short term. Even the long-term models are years, not decades.

So, you've got three really important dimensions of thought that are largely missing from what pops up when you come into work in the morning and you start looking at what's in your inbox and what's on your screen. One easy example of this is the default—if you just do a random stock chart on any financial information provider, it's an intraday chart now that's the default model. Why? It's of no interest whatsoever, but you're seeing that day in and day out, every time you enter something, and it does have an effect no matter how independent-minded that you think you are.

. . . when you're in that state of fear, the last thing you want to do is entertain a new idea or a new point of view, particularly one that maybe threatens where you're starting from. It makes a brittle system even more brittle.

I would say all of these blinders together tend to pull us towards cleverness and away from wisdom, you know. So, we're looking at the leaves on the trees and it's very hard to get to the roots of the more interesting questions. If you get to the roots, that's where the biggest risks are. It's also where the biggest opportunities are. You can make the most money if you get an insight that's closer down to the roots.

The second element relates to where Michael left off, this allocation versus formation, which has very close parallels to exploitation–exploration. Almost all of the tools that I've grown up with in finance have to do with allocation, maybe ninety percent or so. About ten percent of the time and attention is spent on these questions of formation and exploration. There are two elements here: one is that the tools of allocation are handy and theoretically beautiful, as we've discussed. But they really reinforce this mentality that for me to win, you have to lose.

MECHANISMS OF MARKET EFFICIENCY

Michael Mauboussin

Complexity economics is a very powerful way to think about markets in general and the stock market in particular. While practitioners have made astute observations about markets for centuries, viewing the market as a complex adaptive system provides us with better language to describe—and, hopefully, understand—what's going on. The key question I seek to answer is whether stock markets are efficient.

Because SFI has been influenced by physicists, it is worth noting that the term "efficiency" comes from physics. In physics, efficiency measures how energy translates into useful work; in markets, efficiency measures how information translates into stock prices. A perfectly efficient market is one in which all information is accurately reflected in asset prices.

Andrei Shleifer, a professor at Harvard Business School, has a wonderful little book called *Inefficient Markets* that describes three classic ways to get to market efficiency. As a side note, Gene Fama and Bob Shiller both won the Nobel Prize in Economic Sciences in 2013 for their empirical analysis on asset prices. But they are in opposite camps: Fama believes in efficient markets and Shiller believes in inefficient markets. If you are a fundamental investor trying to earn returns higher than those of the market, you have to believe both of them are right, just not at the same time!

The first way to efficiency is by assuming investors perfectly follow the mean-variance approach, which links risk (variance) and reward (mean return) in a linear fashion. You have to assume that investors are rational agents who understand their stable preferences, know the future, and agree on the distribution of price changes. No one believes this to be perfectly true, but it is beautiful. It is a Platonic ideal of sorts.

Mean-variance makes a prescriptive recommendation that can be helpful. It says that there's no use trying to beat the market, and that buying an index fund, which typically tracks the performance of the market, is your best bet. This is indeed sound advice. However, for the strategy to work, there have to be some active managers to take information and reflect it in prices.

The second path to efficiency is based on the absence of arbitrage. This theory says that there is a subset of investors—arbitrageurs—who cruise markets and buy what's cheap and sell what's dear and leave accurate stock prices in their wake. Now we only need a subset of people to be rational. These people really do exist and they serve this function. But what we also know is that, episodically, these people fail to show up. Either they don't have any access to capital or the cost for them to do their job is too high.

The final way to get to market efficiency is thinking of markets as complex adaptive systems. The model of the wisdom of crowds, as described by James Surowiecki in his book of the same name, captures much of what we are after. When certain conditions are met, the wisdom of crowds converges on the equilibrium solution as specified in the mean-variance model.

Let's disassemble the elements of complex adaptive systems. Complex means there are a lot of diverse agents that interact with one another. Adaptive means those agents have rules about how to deal with the world. Learning enables agents to change those rules to reflect the changes in the world. System means that the whole is greater than the sum of the parts. Simple aggregation does not work as the global system emerges from the interaction of agents.

Markets tend to be efficient when three conditions prevail. The first is diverse agents. In markets that means people who have different information, time horizons, and rules of behavior. This explains why behavioral biases are not that important. You can have a lot of little mistakes but the market mechanism deals with them effectively. Second is an aggregation mechanism, a way to bring that information into one place. Double-auction markets, where buyers bid a price and sellers offer a price, are extraordinarily good at this function. The final condition is properly functioning incentives—rewards for being right and penalties for being wrong. For the stock market, this is measured in money.

We know that markets are hard to beat, which suggests that these conditions are generally satisfied. We also know that markets go haywire periodically. Markets become inefficient when one of these conditions is violated, and the most likely condition is almost always lack of diversity. We are social animals. Instead of believing different things, we correlate our beliefs. We all come to the same view about something. The dot-com bubble is one of many examples.

Seeing the world through this lens, versus traditional financial economics, provides a lot of insight. It explains what we see and also explains why we veer to extremes from time to time. Asset prices are a signal, but measuring diversity remains a thorny problem. I think this is a fertile line of research, and it will be informed by new complexity economics.

There's one aspect of diversity that deserves special attention. Blake LeBaron, who directed the economics program at SFI years ago and has been a long-time friend of the Institute, has illustrated this point through his agent-based models. Blake created agents in silico, taught them finance theory, gave them diverse decision rules for how to trade, and let them loose. He knew the value of the asset they were trading. His results replicated the empirical features in the market, including the fat tails.

He shares a figure that shows the asset price in the upper panel and a measure of investor diversity in the bottom panel. The beauty of this model is that you can see under the hood and understand the relationship between diversity and the asset price.

One stretch is particularly instructive. The asset price is going up while the diversity is going down. The agents are adopting the same decision rules and the asset price initially goes up, reinforcing that things are good. But at a certain point, as diversity troughs, the asset price plunges.

Here's the point: the relationship between diversity and asset price is a nonlinear function. This is true in other ecosystems. It's easy to get caught in the trap of a diversity breakdown because the initial feedback is positive—the asset price is going up—until it suddenly plunges. In markets, we talk a lot about "crowded trades" where everyone is doing the same thing. But we have poor measures of diversity.

Complexity economics touches, and informs, many aspects of business and finance. Stock markets, which have a global market capitalization of $80 trillion, are a great example. Neoclassical finance has extended many of the concepts from economics, including notions of efficiency and equilibrium. Behavioral finance has pointed out the discrepancies between these theories and reality. Complexity economics reconciles the two by shedding light on the mechanisms of efficiency and why those mechanisms often fail to deliver idealized outcomes.

There are some really pernicious ripple effects of that as it percolates out into action and into how decisions are made.

The second is, if you ask any real investor, including lots and lots of folks in this room, the money is made on the formation side. The money is not made from shuffling the shells around from one area to the other. The money is made by creating something that's actually needed and worthy in the world. And so the idea that almost all of our tools are not that is a really interesting gap to explore. This percolates really directly over to my work when we think about how to map sustainability, which I'll talk about in a minute.

The third, which is a tough one to talk about, is the role of fear. A lot of times, we brought up fear, we brought up greed, we brought up envy—I would argue greed and envy are just specialized forms of fear. It's uncomfortable for me even to spend thirty-eight seconds talking about this, so that's kind of an indicator of where we are. From the very beginning when Eric was talking I thought, what if when we had that physics envy moment, we just had a better capacity for handling that? What if instead of going down this one path that has gotten more and more narrow, we actually were able to deal with that better and open up all those paths?

That's a really root-level question that I think all of our elegant models are still not so comfortable in grappling with. And that has some important links to diversity. You know, when you're in that state of fear, the last thing you want to do is entertain a new idea or a new point of view, particularly one that maybe threatens where you're starting from. It makes a brittle system even more brittle. And I'll end by just giving you an example of how this is playing out in the sustainability world. How many of you know that ESG is an acronym for sustainability? Pretty many?

Okay, so it's environmental, social, and governance. I spend about half my day talking about how happy I am that there's more data on environmental, social, and governance issues, and then I spend about half my day in deep despair at how this data is being used. The really strong temptation for us is to say, "Hey, we have these big, huge questions that don't fit into these narrow, standard, blinder models that we've created. I'm going to make up a set of numbers that I can shove into the exact same machine as before and crank out a slightly different answer." And that's not the point. We're trying to ask a really fundamental set of questions here. I worry that we have a chance to do something really revolutionary in thinking about true value and true price and true profitability and creating a lot more value for the world, and instead we're going to just bureaucratize our way into oblivion.

I didn't mean to end on a bad note. Let me give you a loftier note to end on. One of my favorite investigations of the last few years—I did a lot of reading on Goethe, and he had this quote, "The corpse is not the creature." He was talking about the passion for collecting butterflies during that time, and how you could measure every leg segment, you could categorize every color on the wing, you could have all the butterflies pinned on the wall. But if you had never seen a butterfly flying, you're missing the whole point. And I would posit that many, many of our models and tools for investing and finance are the equivalent of measuring the legs on the butterflies. I'm really happy to have those measurements. But the reason I keep showing up at SFI is that the questions that are surfaced here are trying to figure out flight. It's much, much more important. It's much more inspiring and that's where I think the real wealth is going to come from.

D. VILLANI I'm Dario Villani. I run a machine-learning hedge fund called Duality. I want to start with two things. Number one, I was elated and pleased to see Geoffrey West saying Ken Arrow is no Albert Einstein. He really said it in a way that made me happy.

[LAUGHTER]

I was a physicist in my past life. I squandered my youth trying to explain ITC superconductivity. And I thought that David Pines and Phil Anderson were doing the same. Instead I found out that in 1987, instead of doing that, they were just spending time with a bunch of economists. That was disappointing.

I want to give just a few observations of where we are in terms of the narrow space of finance. There are very important open questions and we have been discussing the success of the theory . . . The reality is that even assessing what success means, in forecasting or understanding systems that are complex or where the signal is very low with respect to noise, is a very hard question to ask.

If I show two track records for the last five years and I ask any of you which one you think is better, I think all of us should have a very hard time answering that question. That's point number one. Point number two, to me, goes back to the fundamental question—it's very hard to believe, if you think of the evolution of physics, that you could understand magnetism if you were never able to solve the equation of the hydrogen atom. The reality is, Schrödinger solved that and it took up until the '50s and '60s to understand systems with impurity, complex systems, collective modes, all the complexity that we are talking about and referencing.

So, fundamental questions, like, "What is the time scale at which we should retain or let go of information when we make forecasts or try to assess the future?" I ask this question of everybody. I'm going to answer it. This is not a test. To be very successful, the correlation between your forecast and what actually happens is three percent or four percent. Imagine how difficult it is to assess the quality of the forecast for Michael or me, if you're really dealing with three or four percent.

Another comment I want to make is, people have been dismissive of machine learning. Sometimes the classical way to be dismissive is to say this was known in the '40s and '50s—I've already made my joke. It's like saying, "What's quantum mechanics? We knew there were particles; now we just don't know where they are."

If you think of the hierarchy that David was talking about, there are the string theorists like Ed Witten . . . They are at the peak, but at some point, in the '60s, people realized that people who made really the biggest changes to physics were people like Phil Anderson, understanding spin glass, understanding complex systems, understanding magnetism, superconductivity, and so on—I think we're at that very stage.

The most interesting field for machine learning is finance. You have to deal with non-stationarity, low signal to noise. You have to address the problem of rare events. You have to understand how rare events don't distort your ability to learn. You need to be aware that bad things can happen without distorting your learning process to accommodate the delusion that you can actually forecast that.

How do you put all of this together? In my opinion, the Phil Anderson, the Nevill Mott, the magnetism of these days is to be able to tackle complexity the way they did with magnetism and superconductivity.

One joke, trying to echo Geoffrey West. While we are all here trying to discuss complexity, there are groups in the world that have achieved amazing breakthroughs that nobody knows about. You know, there is a group in Stony Brook called Renaissance Technologies that, for the last thirty years, they haven't had a down quarter. They've solved amazing issues in what understanding complex systems is all about. And one thing I can tell you—they have hired astronomers, they have hired mathematicians, they have hired physicists, they have even hired theologists. They never even interviewed an economist. Thank you.

[LAUGHTER]

BILL MILLER I'm an investor and former chairman of SFI. I think it was mentioned earlier about that first meeting that's now the volume on *The Economy as an Evolving Complex System*. And what's interesting with that is the theory in practice; that meeting was held because of the failure of theory, the failure of models.

"The function of markets is to take money away from stupid people who don't deserve to have it." That's Shubik's law.

John Reed had been very frustrated with the fact that none of the bank's models could predict anything and he thought maybe a different approach would help and be worthwhile. Somebody, I feel it was maybe Rob, quoted Paul Samuelson, in my opinion whose *Foundations of Economic Analysis* did enormous harm to the profession. Nonetheless, the quip that he had, which is that the market has predicted nine of the last five recessions—what's interesting about that is that no Fed and no council of economic advisors has ever predicted a recession. So, the market's predictions are much better than the Fed.

And somebody offered the idea that the four that didn't happen, it's because basically the Fed changed their policies to make sure that they didn't happen. I can come back to that later perhaps. But the interesting point about Samuelson's quip is that somebody actually went back and did the data. Since World War II, there've been thirteen bear markets and they were followed by the seven recessions. So, Samuelson had a thing about fifty-five percent—probably the actual number was fifty-seven percent. So that was a pretty good guess.

I wasn't going to bring this up, but since Matt raised this issue of problems of markets, failures of markets, it reminds me of a story. In fact, I'll just reprise the story. When the second meeting of "The Economy as an Evolving Complex System" was held, Martin Shubik of Yale, great game theorist, was in the audience, along with Ken Arrow. Scott Page was then a young guy at Caltech, and he was talking about markets.

[LAUGHTER]

He was talking about markets and about how they did allocative efficiency, but there are things that markets don't do very well. Like, they don't allocate things fairly, so there's a gap between value creation and value distribution. Martin interrupted Scott at that point and he said, "Scott, I can see that you do not understand the function of markets." And Scott said, "Well, what do you mean?" And he says, "The function of markets is to take money away from stupid people who don't deserve to have it." That's Shubik's law, you know. I thought it was an interesting thing.

But if you come at this issue of models and the way things work, somebody had mentioned—I think Eric—about verisimilitude and about Milton Friedman's kind of as-if stuff. But the basic issue between theory and practice is that the models just aren't very good. One of the stories I like about theory and practice is Charlie Munger, Warren Buffett's partner, tells a story of kids in a third-grade class, they're learning arithmetic. And the teacher says, "There are nine sheep in a pen and one sheep runs out of the pen. How many sheep are left?"

So, you just raise your hand and say, "Eight." She says, "Yes." And then a little kid raises his hand and he's a farmer's son and he says, "No, that's not right." And she says, "What do you think the answer is?" He says, "The answer is none." And she says, "I can see you don't understand arithmetic." He says, "I can see you don't understand sheep."

[LAUGHTER]

Let me switch now to sort of the practical side of this, the application side. The most common question I get from people is, is this stuff actually useful, in terms of actual practical application in your job? My comment was, it had been expected to be, but it was very useful. I'll give you three quick examples and I'll be done.

Brian Arthur's work on lock-in and path dependence in the economy was what got us to take a look at technology in 1995 and ultimately buy Dell, AOL, Nokia, and Apple when they were trading cheaper than most other stocks on the stock exchange. He had shown the flaw in Warren Buffett's reasoning about how technologies change so rapidly that you can't invest in them. The point is that underlying technologies change, but, after a point, technology market shares don't change and so they're highly predictable.

Second, a meeting that many of you here attended, "Innovation and Evolution," led us to buy Google on the IPO. It was directly related to that. Since I'm running out of time, I won't tell you how Jim Rutt said he was going to short Google on the IPO because it violated the fundamental rules of economics.

[LAUGHTER]

Then we also did a thing out here called "Money: Past, Present, and Future," and it was about Bitcoin. I don't think you liked Bitcoin then, either, did you, Geoffrey? In any case, that was a thing that was very effective in my opinion because it got me to buy Bitcoin at two hundred dollars. And so the Miller campus—that's the campus that Bitcoin built. There's a direct advantage there.

Lastly, just about a month ago I was invited out to talk to Jeff Bezos and the senior team at Amazon about long-term strategy. One of the questions he wanted to address was, we've got 900 billion market cap and 800,000 employees and hundreds of billions of dollars of revenue and we've been able to scale very well. But do you have any thoughts on scale? So, I got my trusty Geoff West *Scale* book out, dog-eared, went to the section on companies, gave him all the data on that. And he

UNDERSTANDING A DELICATE MACHINE

Bill Miller

The topic of my panel at this meeting was the practical applications, if any, of complexity economics in general and of our firm's experience at SFI. Did it help our thinking about competing in capital markets, or our understanding or beliefs about those markets? Did we behave differently as a result? Most importantly, did we make any excess returns that we otherwise would have missed but for our interaction with the scientists at SFI? In our case, the answer is an unequivocal yes, both as to how we cognized capital markets and how we behaved in them, as well as the tangible results we saw from our time at SFI.

I came out to SFI in the early 1990s, having learned about it from a *New York Times* article in 1987 shortly after the stock market crash late that year. Our firm became the second member of the newly formed business network. John Reed, then CEO of Citibank, had been frustrated with his economists' inability to predict much of anything and so was funding SFI's early efforts in analyzing the economy as a complex adaptive system instead of, as was then and still is done today, as a dynamic generalized stochastic equilibrium (DSGE) system. The key difference between the two is that the real economy is a tightly coupled, highly interdependent system, where small changes or perturbations can have outsized consequences. In a DSGE model there are no booms or busts or crashes; there is not even a banking system. No surprise that DSGE models are of little use to practitioners who have to operate in a messy world characterized by what economist John Maynard Keynes called "irreducible uncertainty."

The work at SFI gave us a new way of thinking about and understanding the capital markets that was much more realistic and useful than the stagnant academic approaches that are still regnant. The great economist Paul Samuelson once quipped that the stock market had predicted nine of the last five recessions. That is much better than the records of the Council of Economic Advisers to the president or the Federal Reserve, neither of which predicted any of the recessions in the post-war period. The stock market is a complex adaptive system that cannot be analyzed in the context of a DSGE model.

The work at SFI gave us a new way of thinking about and understanding the capital markets that was much more realistic and useful than the stagnant academic approaches that are still regnant.

Particularly helpful has been the work of Brian Arthur, who was a resident faculty member at SFI in the 1990s. Brian's work on lock-in and path dependence in technology proved especially profitable because it showed that, contrary to standard belief, the market shares of many technology companies could be predicted with great accuracy, even if the underlying changes in technology changed frequently. We were early in investing in the technology boom of the mid- to late-1990s and benefited greatly from his insights.

More recently, the work of SFI scholars Scott Page and John Geanakoplos helped us understand the methodology behind models that are typically used in finance that attempt to ascertain the value of businesses or manage risk. John's work has illuminated how bubbles in markets can form naturally, thus leading to crashes. The role of belief formation—for instance, how fear can drastically shorten time horizons—has also been incredibly helpful. Just as the novel coronavirus is quite contagious and spreads rapidly, so too with fear; confidence, in contrast, returns only slowly after crises such as in 2008. That has important implications for portfolio construction and stock selection. Also, a 2015 SFI workshop called "Money, Past, Present, and Future" explored cryptocurrencies such as Bitcoin in a much more systematic way than had been done before. Economists, anthropologists, evolutionary biologists, and cryptographers contributed insights from their particular fields that were crucial in our decision to invest in Bitcoin when mainstream finance practitioners denounced it as a "fraud" or an instrument mainly of use to criminals. Its price has risen around thirty times since that workshop.

In an article called "The Great Slump of 1930," Keynes, who was both a brilliant economist and a great investor, said, "We have involved ourselves in a colossal muddle, having blundered in the control of a delicate machine, the working of which we do not understand." SFI is contributing to greater understanding of the delicate machine, the economy, so that our blunders may be more infrequent and less costly.

interrupted—we're about halfway through the thing and he says, "That comes from Geoff West's book, right?" I said yes. And he said, "That's a great book." So that's my intro here.

P. J. DAVIES I really want to build on that last part of what Bill was saying. The interesting thing about Eugene Fama and the efficient market stuff coming out was that it spawned an entire industry of money managers. It might not have been true, but it was incredibly useful. And it was put to work, and the stewards being put to work now I guess with complexity economics and agent-based modeling and that sort of stuff. I mean, it is being used, but more in ways of testing changes in policy, testing new products, looking for vulnerabilities in systems.

It seems like much more of a risk-management thing in many of the things that I've seen so far. So, I guess my question, to put it deliberately, crassly, and only semi-seriously is, in what ways can it be applied directly to trading like this? Or, how can complexity economists get rich?

K. COLLINS Well, I'll start. Just to reiterate a premise I hold, which others strongly disagree with—I mean, all my friends who are wedded to the Fama approach would strongly disagree with this. With every level of abstraction you make, your models get prettier, but your link back to the actual world is further and further away. And so, for a fundamental investor, one thing that's really difficult with some of the tools that have been developed the last twenty years or so is they're taking you further and further, in my case, from anything I actually care about.

I strongly believe if you're a real investor and not just a trader or a financier, that you're trying to create something of worth in the world and be a small part of that by providing the capital to do so—if that's your premise, the more abstraction, the harder it is actually to fulfill your premise. There's a real place for those models but it has helped me a lot to look at finance as kind of a whole ecosystem unto itself: where we've seen the greatest growth, where things might be out of balance. The more synthesized parts of finance have vastly outscaled the more connected parts of investing and that's inherently a precarious place to be. I find that most of us aren't even conscious of that day in and day out and it's helpful to be so.

D. VILLANI One comment. Personally I believe you need "AI" or machine learning and know that you should part ways completely with top-down modeling. It's because there is a fundamental difference in finance. The act of learning something about the market dynamics of deploying capital to it changes the dynamics of the market itself. When you recognize a cat, you don't change the nature of the cat. (It's always the cat for artificial intelligence. I don't know why.)

But, if Michael and I notice something completely spurious—say, every time the twenty-day moving average crosses the fifty-day moving average, it's a buy. And we have a hundred billion dollars each and we convince ourselves to start deploying capital to it, that noise can actually become a signal. A signal can become noise, too. So there is reflexivity, the positive and negative feedback loop. What it means: you need a structure that allows you to transition across qualitatively different models, every relationship. There is no Newtonian law of markets; they are all ephemeral relationships in a sea of noise and the only way you can do that and capture non-linearity and complexity is with a system that is rich enough to be able to contain all the models, so a universal approximator—that's what neural nets are—and allows you to do that.

The next part, and it's not necessarily related to neural nets, is how you dynamically evolve the coefficients or the distribution to time. And that's a very complex problem. It's a problem that scales exponentially if you don't do it carefully. You can do better and so on and so forth. But the reality is, given this negative and positive feedback loop, there is no way any model can sustain itself. Like economists say, "Once a measure becomes a target, it stops being a good measure." So, the act of recognizing anything changes the market dynamics. That's very special to finance.

B. MILLER I'd just say that the stock market has many different functions, but one of the ways it operates is as a real-time, information-processing mechanism. Information is created, stuff's going on, it's incorporating those into prices as they happen. So, what's the probability that IBM will be bankrupt tomorrow? Rounds to zero. How about three years from now? How about ten years from now? How about twenty? As you move out in the time horizons, what happens is, the code of uncertainty widens about what can happen. We like to say that more things can happen and will happen. Then you can try and exploit some of these anomalies, like the fact that, when people are under stress, their time horizons shrink.

Part of it right now is what I'll just call "time arbitrage." There are more opportunities out there the longer your time horizon is, and when you're competing short time horizons with Dario, we don't have any competitive advantage; most people wouldn't, either. But we do think we may have a competitive advantage structurally, because most firms have measurement periods that are much shorter than three to five years. There are ways in which I think people who understand, broadly speaking, complexity economics, which I think is really just kind of about insights that are more closely aligned with some of the aspects of the market in being less—not non—but less model dependent.

. . . once a measure becomes a target, it stops being a good measure. So, the act of recognizing anything changes the market dynamics. That's what's special to finance.

M. MAUBOUSSIN We did a thing on complexity in commerce in September in San Francisco, and I saw Brian. I said to him, it's remarkable that when he first—or at least I first—learned about a lot of stuff on increasing returns and network effects and so forth in the mid-1990s, that was very novel, and certainly not mainstream. And then it went and became much more popular and now it's overused. So I think a deeper understanding of how these things work itself can provide a lot of insight. And I'll mention something that's somewhat tangential—that I've personally been incredibly influenced by the work of Scott Page and others here on how to think about organizations and, in particular, how to think about the role of cognitive diversity in identifying different alternatives for the future. And that's one thing that I think, as investors, we really struggle with; we think that we understand the future better than we do. So, opening up minds in that way I think has also been incredibly helpful.

P. J. DAVIES To expand on that a bit, one of the things we were talking about there is having very specific performance indicators, very specific numbers that you hit, and having inflation targeting and monitoring GDP. These are numbers that we can achieve. These are things that we can recognize and measure, but we get to ignore a whole bunch of other stuff while we're doing that. And I guess this is particularly relevant to your work, Katherine.

With ESG in particular, there are a lot of new targets, new measures being set. They are very specific and can ultimately become very limiting and subject to Goodhart's law as well in the way that you've mentioned, Dario. How can we use complexity economics to expand the frame of reference and then to ensure that we get a much better outcome?

K. COLLINS It's a big question. One thing that I've really benefited from in my time hanging out at SFI is, I've gotten a lot more comfortable saying, "I don't know." I don't know because it's not knowable. If you say you know, you're wrong. That's not culturally endemic to most financial organizations, and so one thing I'm noticing, as more folks are concerned about sustainability, again, it gets back to this root of fear, "Oh my God, this is really awful. Let's fix it. Let's fix it fast." Instead of taking the time to say: What do we know? What don't we know? What would we love to know that we have only the faintest clues about right now?

We're sort of rushing ahead to say, I need a metric for this. I need a metric for this. I need a metric for this. And as soon as the metrics are created, I'm going to have a score. And then as soon as a score is created, I'm going to re-optimize my portfolio on it. You're really far now from the original intention, but it feels really good. It's very highly satisfying and it fits really nicely with the more efficient elements of the machine of investing and finance. And so the help that places like SFI can provide I think is in reminding us of what a broader ecosystem model looks like. To Bill's point, they can help extend our time horizons. There's still a big dispute about whether or not environmental, social, governance concerns are financial returns. If you extend your time horizon long enough, yes. In fact, that requirement is shrinking by the minute as certain circumstances become more dire in the world.

And so, we have finally come to accept the idea that these are relevant investment issues. But skipping ahead to having this over-automated solution really disturbs me. For example, I've talked with a few of you this weekend about the important of "S" in your organizations. Intellectual capital, employee engagement, customer loyalty—you know, if I'm trying to measure diversity at a firm, what are the metrics they have right now? Number of women on the board, and maybe a little bit of diversity within the C-suite. That is not my question. I'm happy to have that data and those are important issues, but they're so far from answering the question: is this an intellectually robust environment where the company is actually able to adapt to changing circumstances over time? That mismatch is really essential and it's something that the work of SFI and everyone in related fields has helped to illuminate.

D. VILLANI I'll give a much narrower answer. In machine learning, for example, there are standard tests of data where people check the performance of whatever is their tool, can be recurring neural nets and recognizing images, recognizing characters. I think that people can devise—we do—what targets or experiments would assess the value of adding complexity and having complex economics. A lot of people, for example, ask, "Why do you use neural nets?" Because it captures no linear effects. "Are linear effects important?"

"Does it improve your forecast? How does it improve it?" These are fundamental questions. A lot of the modeling when you do physics is really to transform from extensive to intensive variables. You discuss volatilities and you don't discuss prices of options. You should discuss the covariance structure of your forecast to what actually happens; you should not discuss what is the hit ratio of a strategy over a period of time. Really getting at the core of what makes a methodology superior to another is one of the open questions that people should be able to address and they don't address, actually.

It is interesting to me how long some misunderstandings can last. Everybody, there is a trillion dollars managed with factors: the Fama-French. People always talk about how "Momentum has done really badly." Like there is this entity that people talk about to build narratives. We did this experiment: we generated factors returns. From there we computed stock returns. And from there, using the techniques that everybody uses with some cross-sectional regression, we computed factors returns. Guess what? You don't get them.

You can actually observe sometimes the factor returns went down and the way people (conventionally) compute it, it went up. There is a whole narrative. There is a trillion dollars managed by people who still don't address collinearity, they don't do filtering, they don't use Kalman or particle filter . . . it's very rudimentary and with a trillion dollars, people writing books, even people getting Nobel Prizes when, actually, the methodology doesn't recall the artificial factors you can generate. So, there are fundamental questions that really need to be answered.

P. J. DAVIES Another thing I wanted to ask you about is the character of agents in financial markets now. You kind of alluded to it there slightly, Dario, but especially in the US stock market, it's incredibly efficient in many ways. Everybody gets more or less the same information down more or less the same pipes at more or less the same speed. There's a lot of trading being done by machines or done passively or done with rule-based systems and done systematically in ways deliberately to extract human emotion and bias out of trading.

I wonder if any of you think that changes how markets ultimately function in a way that makes them . . . perhaps stable is the wrong word, but maybe they exhaust the possibilities quite quickly and then hit sort of stasis. One of the things we've seen in financial markets in recent years is long spells of very low volatility followed by short, sharp shocks. I'm wondering if the makeup of agents has anything to do with that.

D. VILLANI I'll give you a short, one-minute answer to this. First of all, people love to talk about the quality of their forecast and the information, how it's translated in forecast, when everybody in the room would agree that in finance there's low signal to noise. There are two corollaries to that. Corollary number one: we should not talk about deep learning, as I said before. Corollary number two is that our ability to forecast is naturally capped. The difference between me with super-sophisticated, non-linear, complex models or somebody else with a linear model is tiny. What makes you successful is everything else. It's like, how do you build a covariance matrix? Do you remove the random matrix part of it properly? How do you evolve covariance? It's like hedging beta and not squandering that little bit in a silly way of constructing portfolios or executing.

As far as these bouts of volatility and so on . . . definitely, again, "If a measure becomes a target, it stops being a good measure." Everybody tries to be factor neutral, market neutral, this and that . . . ultimately, there is a lot of crowding. You see it during the months in which there is delivering. Everybody gets caught with their pants down. So, the risk management is really an overlay that causes a lot of crowding because they're all managed in the same way. Very much like the crisis of 2007 and '08 when people were arbing ratings.

K. COLLINS *[to Michael]* You've done a lot on this.

M. MAUBOUSSIN First of all, I agree with all of that. I mean, look, active managers play an important role in markets in two regards. One is—and I'm talking about more fundamentals as Dario's doing more systematic stuff, but as price discovery is basically making prices efficient and then on liquidity. On price discovery, the one thing I will mention is there are people who suggest, for instance, that more people indexing makes it easier for active managers. But the way to think about active management, for someone to earn excess return, someone else has earned the same amount of under-performance.

It's like a poker table. The amount of money coming in at the beginning of the night and amount of money going out at the end of the night just gets moved around. The smart people get more and the less get . . . So, my hypothesis is actually that people who've indexed are the least skilled investors, which means the smartest people are competing against one another. That just makes it even more challenging and that would be sort of my conjecture.

I do think that there are a number of issues that we just don't know yet. One is the degree to which all this indexing has affected price discovery and efficiency. There are things like stocks going in and out of indexes—does that affect the valuation, the liquidity, and so forth? And then the other thing is, we have had these liquidity shocks. Dario mentioned this thing with these big, huge multi-strat firms; all are doing very similar things with very similar risk management parameters.

They pull the plug at the same time and do the same thing. So they're having an influence. But I think we don't really know about liquidity. Liquidity is the ability to translate an asset into cash or cash into an asset as frictionless as possible. And I think the jury's out. I would not be shocked to see at some point in the next three or five years a really big—you know, we had something in 2016, something else in 2018, but I wouldn't be shocked to see a really big liquidity scare. Again, most of these indexers, myself included, are just sitting on this money. So liquidity might dry out in a way that's quite frightening.

P. J. DAVIES Bill, do you have any thoughts on that? Or maybe I can ask you the same question in a different way: Is it harder to identify value that can be catalyzed, that can be realized by more investors seeing the investment perform in markets that are sort of more governed by a lot of this sort of stuff that you guys just mentioned?

B. MILLER Yeah. I don't know if it's true or not—but I *think* it's true—that it's much harder to outperform short term. By that I mean in any given year, certainly in any given entry year, it's easier to outperform on a multi-year basis. Since the financial crisis, I think we're in the top one percent or two percent of all funds. Part of the reason for that is that if you're forecasting economic variable GDP or auto sales or interest rates, you're just forecasting behavior.

If you think about a line that I think Kahneman and Tversky had in their work on prospect theory, they said that probabilities attach themselves to the descriptions of events and not to events themselves. Which gets to the point about narrative economics. Josh talked earlier about epidemiological models, about how things spread. There's probably a lot to that. But, to give you a very simple example, after the financial crisis and the market kind of starts up at the bottom of March of '09 and we're sitting around saying, "Okay, so what's the impact of this on people's behavior? How long will it take if there's an impact and what is it? If it's significant, how long would it take for it to settle out, for it to be forgotten?"

Part of what we concluded was that it was a shocking event. People didn't know they'd taken that much risk. Their house fell thirty-five percent peak to trough, ten percent unemployment. A lot of fear that certainly engendered a desire for safety and a much greater desire for safe assets. So: bonds, bond proxies, utilities, consumer safest tip stops. But, generalizing that, our team's view was that, basically what we have is probably a big gap between perceived risk and real risk.

People are always going to overestimate risk because they're so terrified of what happened. One of the things that we decided to do is, wherever there was perceived high risk, we'd investigate to see what we thought the real risk was. Generally speaking, if you could have done only one thing after that crisis for ten years, it would have been basically just to buy the high beta names, the ones that are perceived riskier and more volatile in the short run, because those have significantly outperformed most other things. Not everything—utilities and consumer staples have done really well, but those things are stupidly overpriced, in my opinion, as are bonds.

My last comment is that there are going to be some of these persistent anomalies. One of the most amazing ones is the fact that every investor knows that the value of any investment is the present value, the future free cash flows of the investment. Michael pointed me to an article some time ago that ninety-five percent of all analyst reports contain an earnings model or an earnings forecast, and only about five percent contain a cash-flow model or a cash-flow forecast. Even though that's the driver of value, there are all kinds of structural reasons why they think that even though you can look at long-term statistical analysis and see that the relationship of earnings and stock prices—the correlation's very low.

D. VILLANI Can I add something on cash flow, actually? I'll try to play economist a bit. One of the things that we found very interesting was, once you have interest rate at zero, a cash flow in the future becomes as important as a cash flow you have visibility on. The reality is that a cash flow in the future has much bigger error bars around it. So, if in the past when the rates were normal, around seven percent, you say, "I have visibility over the next three or five years," could Google one day disintermediate JP Morgan Chase? Yes. But that scenario, even if I tried to price it, discounted to today, impacts probably ten to fifteen percent of the valuation. It doesn't make sense to spend that much time on it.

Instead, when you have interest rates at zero, those scenarios become hugely important. If anything, fundamentally, the economy becomes much more of a gambling economy, because the future that has much bigger error bars becomes as important as the present where the error bars are much smaller. It's not easy to deal with that.

P. J. DAVIES I'd love to open it up to the floor now for some questions.

AUDIENCE MEMBER One of the avenues where I think complexity economics could help is the concept of algorithmic asset allocation. Dario, you were mentioning Renaissance Technologies. My hypothesis is that they've probably cracked something like that. I have no idea how they make this much money. But the question is, do you think that's one of the things that many of these concepts could be utilized for? Like a better way of aggregating the stock picking?

We've been really good at building robots or algorithms that buy individual stocks. We haven't yet figured out how to put everything together in one portfolio. I'll give you my hypothesis. I'm sure it's wrong, but I would think that it would have to deal with something like an evolutionary algorithm. Because what you want is to live for the next day; don't run out of cash. And probably an objective function, something like prospect theory or something like that where you're bringing into play the human component.

D. VILLANI Yes, I think that's one of the most important issues to solve. And I'm sure actually they've had breakthroughs that would open eyes for everybody. We hired the guy who won the 2018 Fields Medal last year.

But if I can make a joke, here's the following: Markets are not stationary, so imagine there is a god that really loves us. He says, "I'm not going to give you any forecasts, but every day I'll tell you the time scale at which you have to compute your statistics." So today, the god comes to say 200 days. Tomorrow, 150 or 300 days. Imagine—answer this question: you have a portfolio of stocks, a thousand stocks, and a god just told you, you can use only a hundred days. That's the relevant time scale. Do you know how to estimate a covariance matrix with a hundred data points for the thousand stocks in a way that is sensible?

As far as Renaissance, why am I obsessed with it? I interviewed Jim Simons and I said, "Listen, there was an amazing group of people. They were all physicists like Bethe and Fermi, and they were in Los Alamos. In two and a half years they built the atomic bomb. Well, within ten years, ten other countries also had the atomic bomb. How's it possible that you have had breakthroughs for thirty years?"

Yet there are still people having conferences discussing how to build the portfolio. It's like we are in the Amazonian forest with our spears and there are people there building an atomic bomb and we have no idea what to do with it. They've had breakthroughs that have solved all of these issues and still they're discussing if linear effects are important, should we use random matrices. There are still people using factors, Markowitz . . . It's funny that even the atomic bomb took only ten years and here we have somebody making $150 billion on PnL and there are companies with hundreds of billions of dollars they manage. They are not even at the beginning of that path. It's unique, in my opinion.

JOSHUA EPSTEIN Katherine, I was delighted to hear that we share an interest in fear and the epidemiology of fear. One of the things that's interesting about fear learning, the acquisition of fears, is that it's fastest when the stimulus is salient and surprising.

The same stimulus could be fear-inducing or not depending on expectations, so fear learning is expectation-dependent. But the interesting thing about fear—and Bill, you mentioned that stress reduces our time horizon—is all the experimental literature on how fear also distorts our probability estimates. I mean, it interferes with and biases upward our probability of further salient events, leading to a spiral. So, the mechanism of fear acquisition can be self-amplifying, which I think has lots of importance for markets. Michael, I don't know whether that's canceled out by other biases or not, but—

M. MAUBOUSSIN Not if we all have a fear at the same time, right?

J. EPSTEIN Yeah.

K. COLLINS There are a couple of elements here that I think are worth exploring. I'm aware we've got such amazing technical expertise in the room. And so one question is, "What can any of us add to the conversation?" We've had this kind of frustrated discussion in the hallways: "Where's our new model? Where's our new understanding?" It might be a while, is my takeaway from this discussion. Then the question is, "What do you do in the meantime?" I think you might've mentioned this morning—we can't control our fears. But we can definitely control our ability to respond to our fears. We could, at the very least, stop being such big jerks all the time. So then the question is, "What does that actually look like in practice?" Again, there's some very elegant theory that could go with this. There's some very elegant modeling.

I would be much more interested in a model that explained how cooperation takes off, which some folks here are working on, or how affinity—and, dare I say, love—work, because that's really your only chance to cancel out the power that fear has at the end of the day. And so what I'm trying to do with my team . . . I know this sounds very kind of JV and grassroots, but the question is, how can you be ready for times that are going to come and challenge all of the tools that we have at our disposal?

One way to be ready is to practice. So, we practice all kinds of different simulations in a formal way, but we also practice personally, like, "Hey, you were really mad in that meeting. Is there a way that we could actually work better together so we have time to get to the better question, instead of just giving up on the conversation altogether? Maybe getting that much closer to an answer?"

We do things, for example, with my research team. Instead of having them, very first thing, sketch the competitive environment for any company they're looking at, I ask them to draw me a landscape map. I want to see all the cooperation, I want to see all the partnerships, I want to see all the customer dependencies. And then you can layer in the competition, but it's within its proper place. It's not the only thing that we're looking at from the very beginning.

Friday mornings, my team—and this is super hokey, but I'm going to say it, I'm practicing courage—we do a thing, "best things." By Friday, everyone's really tired. We've got a million things on our agenda, we've got a lot of important decisions to make, and we're kind of grumpy. And so, instead of starting from that place of scarcity and gotta get it done and everything's gone wrong this week, everyone has to say, what's the

NO SUCH THING AS AN EXTERNALITY

Katherine Collins

On my very first day of my very first economics class, Professor Julie Matthaei explained externalities—the untidy pieces that don't fit into a model, so we put them in a corner. Fortunately, Julie was also a wise and radical economist, so at some point during every class discussion we would come back to examine this big pile of issues we'd set aside—environmental impact, unpaid labor, and the like. She explained that an externality was not irrelevant, but, rather, uncounted—a consequence without a cost. One of the biggest lessons of my career has its roots in that Econ 101 class: in the real world, there is no such thing as an externality.

Our SFI meeting on Complexity Economics was a dream come true for anyone who is trying to invest in a more holistic way. The meeting centered on a premise that businesses that solve human problems (to paraphrase Eric Beinhocker) are also the ones most likely to create lasting value. I drew three main observations from our discussions.

First, many practices of finance suffer from triple blinders: we use only one unit of measure (usually dollars); we use only short units of time; and we usually rely on equilibrium models that sit just under the surface of our standard calculations. This has the result of creating tidy and sometimes beautiful models, but the narrower and more abstracted these models become, the less they reflect reality.

Second, we spend most of our time in finance focused on allocation and not formation. This is partly because of the modeling challenges involved with examining formation, but, as Dario Villani says, "You should not distort your learning processes to accommodate the delusion." Unfortunately, this same mindset is taking root within the sustainability community, where in our fervor to have comparability and objectivity, we spend endless hours on backwards-looking and incomplete metrics of environmental, social, and governance performance instead of seeking out generative solutions that would produce benefit and profit. Every investor knows that the best investments come from formation—from creating something of value in the world, and yet our tools tether us to something less.

. . . many practices of finance suffer from triple blinders . . .

Third, we are fearful creatures and go to great lengths to preserve a sense of certainty, even when we know it to be false. As Brian Arthur reflected, we would much rather have an ice-palace model, brilliant and brittle, than one that is messy and alive. This is not just an aesthetic or analytical preference; it is because we are so deeply uncomfortable saying, "I don't know" or "it depends," especially in finance. This human element intersects with the issues above in a powerful way, keeping us trapped in a sea of cleverness without the courage to seek more complete wisdom.

In my work, when I'm frustrated by the shortcomings of our simplistic metrics, I often invoke the Goethe quote, "The corpse is not the creature." He was talking about the study of butterflies, how we can measure every leg segment and categorize every wing color, but if we have never seen them flying, we are missing the whole point.

This is why I spend time with SFI. Here, we ask the questions that get us closer to flight.

best thing that happened to you this week? And I gotta say, some weeks the bar is very low. One of my colleagues said, "Well, I got a haircut."

[LAUGHTER]

"Looks pretty good." I mean, we are stretching sometimes to do this, but what does that do? And I don't call it this; I mean, it would freak everyone out. It's a gratitude practice, right? It gets you in the habit of saying, "Look, no matter how crappy this week has been, something good has happened. I can handle this, I can move forward in a thoughtful, resilient way, and extend that time horizon back out to a reasonable opportunity set instead of eliminating it."

RAY IWANOWSKI[1] A couple of you talked about time horizon of investments and signals. I've talked to Michael Mauboussin about some agency problems in the asset management industry that a consideration of complex adaptive systems might help to address. Specifically, consider an individual agent who accumulates wealth over time, earns returns on that wealth over time and consumes the product of that wealth. Theory suggests that he would invest to maximize some measure of expected utility of that consumption over time. However, most of us delegate investment decisions and the ongoing assessments of risk return trade-offs to pension fund managers, mutual fund managers, or hedge fund managers, who then have to answer to other agents such as allocators, consultants, investment committees, boards of trustees, etc. So, if the ultimate objective is to maximize the individual agent's expected utility, but agents act as delegates in driving decisions around that objective, there may be a disconnect between that agent's risk tolerance and time horizon and the utility and objectives of the delegate agents along the way.

To be more concrete, I may have a very long time horizon for my investments and may be willing to take a fair amount of risk over short horizons to provide higher returns in the future. My delegate agents, on the other hand, may be incentivized to consider a much shorter time horizon and much lower risk tolerance since their objectives may be around retaining their assets under management and/or their jobs.

1 SFI ACtioN member and Managing Principal/CIO, SECOR Asset Management.

So, Bill, I'm jealous of you if your clients do give you a long horizon, or Katherine, if you're doing sustainability investing, maybe people derive some utility from the fact that their investments are supporting sustainability, irrespective of short term returns. And perhaps over five to ten years, there is a high likelihood that such confidence in your process will be vindicated as strong returns come to fruition. However, for many money managers there is a high likelihood that they will lose their mandates if there is poor performance over a year or two, in some cases even less.

I believe that these stylized facts of the investing community and markets could lead to some "perversion" of strategy performance, decision-making and ultimately asset prices. I think some of these issues and possibly solutions can be well represented in a model of complex adaptive systems. For example, how do these agents of the delegated asset-management industry feed into investors reacting too quickly to drawdowns when there's no practical, good reason to have to sell into the storm, especially if the positions are unleveraged? Such dynamics can explain why decisions are often driven by short-term fear—i.e., the fear that the strategy may not rebound quickly enough before clients liquidate, even though the ultimate individual agent does not care about such short-term fluctuations. So, I think that might explain a lot of this disconnect on time horizons. Is that an interesting way to understand markets? I would love to know your thoughts.

M. MAUBOUSSIN I've written a little bit about this. Everyone knows about principal–agent problems and I call it the *principal–agent–agent–agent problem*. The way you might think about this, think about your endowment at the University of Chicago or whatever it is; that endowment's going to have a board, often an investment board, which is then going to hire a consultant, which then hires a money manager. So, you have the ultimate beneficiary of this endowment. But you have these three agents in between. And, by the way, they all have to justify their existence on some level. Even the investment community may be getting paid, but they feel like they have to do something. These consultants have to do something, and the money manager, and so that can lead to a lot of bad behaviors. There's been just a ton of research done on this about firing and hiring managers, and moving asset classes, and so forth. And, by the way, there's an amazing story in Chicago around the financial crisis. It just leads to all sorts of slippage, right?

There's no question that that's a huge factor. Just in thinking about asset prices in general, one of the things that gets short shrift in the discussion is flows: money going into funds, money going out of funds, and how that affects the performance of those funds themselves.

D. VILLANI I think that there are also the difficulties, as I said at the beginning, of assessing the quality of your forecast. People generally talk about Sharpe ratio. They talk things that are very dependent on only one realization of the many possible paths. People don't talk about the quality or the covariance structure of your forecast to what actually happens.

So, people are flying blind. Of course they're going to assign huge error bars. They say, "Trust me, in ten years it's going to be okay." So, the discussion doesn't happen at the level where it should happen. It's anecdotal, it's all "trust me," and the measures that people use are just silly, and they're very much vulnerable to one statistical path. They mean absolutely nothing.

AUDIENCE MEMBER I have one comment to Dario. I agree with you about the zero interest rates. But now, think about negative interest rates. If you think something's worth a lot more in the future because interest rates are zero now, it's actually more valuable in the future if interest rates are negative. So, that's one point.

Actually, I have a question for Katherine. I have been a value investor for twenty-two years, probably longer, and I'm still confused about ESG versus SRI.[2] Let me give you a real example: I run separate accounts. So, we email the clients and ask them, please tell us which stocks not to buy, like tobacco stocks, cigarette stocks, oil stocks—which ones fall into your ethical dilemma box? And, if you go through the answers, the answers basically fall under political lines. If you're a Democrat, you're not going to buy guns, and vice versa.

So, if I ran in social money in just one mutual fund and I was forced to make socially responsible investments, it would be impossible, right? That is what I wanted to ask you. How does one actually do this?

2 Socially Responsible Investing.

K. COLLINS Yeah, we haven't made it easy as the field has evolved. The roots of sustainable investing go back even before SRI, but the earliest question was, "What are you against?" That, basically, was the starting point. What are you mad about, essentially? What do you want to divest from? Everyone here who studies theory—you're automatically ready to fight; you're shrinking your investment universe!

We have finally started to move into a much more interesting set of questions, and I think this is a vital set of questions, not just for me and my portfolios, and not just for investment practice, but for the outcomes of how finance affects the world. A much more interesting question is, "What are you *for*?" What do you want to get your money behind in the world? What's interesting is, when you ask that question, three things happen. One is, you get massive convergence that is totally non-politicized. There's no one who doesn't want clean water; there is no one who wants their children to be sicker than they were growing up. I mean, it's a really quick and rapid convergence.

The second thing is, you do get this extension in time horizons because you get away from whatever is in the news today that is objectionable or causing a lot of controversy to, like, what you really want to see in the world over time for your children, for your grandchildren.

And the third thing is, you move from this space of constriction—again, fear comes into it. I can hear my Chicago friend yelling in my ear as soon as someone says, "exclusionary approach." You get into this question: What do you want to include in your decision-making? What do you want to include in your portfolio? It's inherently a much more generative space. To translate this to a fundamental context, if I sit down with a CEO, as many of my colleagues with similar titles do, and I pull out a big report card from a third-party analyst who they don't know or respect, and there are eighty pages of information there, fifty of which are irrelevant, twenty of which are inaccurate, and ten of which are actually pretty interesting, and I start going through page by page and taking points off this report card, you can imagine how useful and positive of a conversation that is.

If, instead, I say to the CEO, "What is the greatest thing your company is doing? What is the greatest thing you could do?"—everything opens up! You're talking about creating something great. And that is, I think, underestimated in every possible dimension. It's got amazing potential for financial returns, amazing potential for positive impact, amazing potential for long-term capital allocation, and you've skipped right over the let's-stay-with-all-the-dysfunction-that-we-have-right-now. It orients everything towards a much more positive placing. Not in a Pollyanna way, but in a "Look, if you are dissatisfied with where we are now in any dimension, the answer is not going to come from fighting more and more about that. The answer's going to come from creating something better."

I really think this is an essential shift. And again, that's one reason I'm worried about some of the common practice, such as going down the report card framework, which I think is a bad investment idea and bad for impact. It's a lose–lose, and has worse social and environmental outcomes, potentially, and also worse investment outcomes and creates a lot more bureaucracy.

> "What are you *for*?" What do you want to get your money behind in the world?

D. VILLANI Do you think a socially responsible fund is one that is very tax inefficient?

K. COLLINS No. Why?

D. VILLANI Let's say I create a structure that is the most inefficient possible, so society's going to get more taxes, right?

[KATHERINE LAUGHS]

K. COLLINS That's kind of funny! I haven't gotten that question before. You'll notice I don't use the language "socially responsible," partly because it's very divisive language. If you say, "This is my socially responsible fund," you're basically implying everything else is not, which is inherently a very bad place to start, right? So it leads you down a tough path.

M. MAUBOUSSIN We gotta get to Jim Rutt. Jim, you ready to rock?

JIM RUTT[3] Yeah, I have a question for Bill. I definitely agree with you about the oddity of the fixation of the investment community on "GAP earnings." I remember back in my days as a CEO, I always made sure I put a big chunk of my quarterly call—"Yeah, we've got some good earnings here, but the story is really cash flow!" In your mind, what's a convenient way to systematically invest on "arbing"—essentially, companies that are differentially valued for their earnings versus their cash flows?

B. MILLER Sure. Actually, for those of you that don't know, Jim had probably the best timing of anyone who sold his company. I think you sold it on the exact high, right?

J. RUTT March 10, 2000. That's best, to my calculation, within eight hours, that was the top.

B. MILLER Yeah, I think you told me that when you became CEO, you figured, maybe if you did a great job, in fifteen years or so, it could be worth five billion. It was worth five billion in, like, six months, or something like that.

J. RUTT Yeah, I figured eight billion in five years. And within six or seven months it was twelve billion, and I said, "Sh—t." Time to sell!

B. MILLER Probably the best example of a misunderstanding of this valuation of anything has been Amazon over its entire history. It was constantly, you know, Baron saying, "Amazon.bomb," and everybody talking about how they don't make any money, they don't make any money, they don't make any money. CBC did a special on it, and one of the things they interviewed me about was Amazon's finances. They start off with, oh, how can you own this thing that doesn't make any money? And I have two answers to that.

Number one, John Malone, the great cable investor. I said, if you put one dollar in John's TCI when John became the CEO, and you kept it there for the twenty-five years he ran that company, that one dollar is 900 dollars. And he never reported a profit in twenty-five years. So, something else is going on besides profits and the report of profits.

---

3 SFI Trustee.

With Amazon, I said, look, there's a reason that they're called "generally accepted accounting principles," and not "divinely inspired accounting principles" or "immaculately conceived accounting principles." They're a way to capture a certain type of information for particular kinds of companies that have a particular kind of economics.

Michael's written about this forever. It's just all about looking at free cash flow, and the growth of free cash flow with capital allocation. Someone else may know something different, but . . . I've been doing this almost forty years and the only company that I have ever seen, that, when they report their quarterly "earnings," don't report the earnings—the first thing they report is their cash flows—that's Amazon. They report their cash flows, their free cash flows, their operating cash . . . and way down, then they talk about what their gap earnings were. I think just getting that right, getting companies to focus on that and, again, just focus on the actual underlying economics and not how those economics are reported.

M. MAUBOUSSIN Can I add something really quick to that? There's real-issue paper, which I'll share with you if you like. So, the last thirty years, we've had a watershed change where most investments now are intangible rather than tangible investments. Right? That's a big issue. What these guys did is, they laid out a framework for thinking about what we'll call "Operating SG&A" versus "Investment SG&A." They're saying, let's take the investment, we'll segregate it, we'll capitalize it on the balance sheet, whatever it is. When you do that, what happens is your earnings get a lot better, and your balance sheet gets like . . . it goes back to Bill's point.

The free cash flows are unperturbed in all this. If you're focused on free cash flow, you never miss this beat. But it allows you to understand what's going on. And, of course, you then know the magnitude of the investment; you can assess the quality of the investments and likely payoffs. So, there are these techniques. But, again, free cash flow is obviously the lifeblood of value.

J. RUTT Yeah, it's the one you can't fake, also. The earnings, allegedly—and we probably do a better job of standardizing than we used to—but still, they're based on a compounded set of fictions. Cash flow, can't fake.

COMPLEXITY: POETRY FOR THE MILLENNIUM

Dario Villani

So much has changed in the world since SFI's New Complexity Economics symposium last autumn. Paul Davies introduced our panel by recalling the 2008 financial crisis and commenting on how hard it was back then to find a contrarian voice that challenged the view that the "go-go days" would never end. In exchanging perspectives on the topic of complexity in that beautiful setting last year, no one had any idea a pandemic was about to alter our experience as humans, and the way we interact and work with one another.

Predicting outcomes to events with no historical precedent is nearly impossible, which is why it's so important for investors to embrace a methodology that respects complexity. Your strategy has to accommodate novelty not just by recalibrating parameters, but by transitioning across qualitatively different models. That kind of intelligence requires an open mind and acceptance that the forces driving markets at any point in time can't be boiled down to just a few factors or "known effects," capable of being understood and captured with top-down modelling. This intelligence has to be able to adapt to what matters to markets locally, living on a lower dimensional manifold which evolves through time in the space of all models.

Earlier this year, Kharen Musaelian, my partner at Duality, co-authored a paper with Alessio Figalli, a 2018 Fields Medalist, entitled "A Scale Dependent Notion of Effective Dimensionality." The paper shows that the effective dimensionality of a neural net is much lower in rank than the headline number of parameters might suggest, connecting neural nets, information theory, and fractal dimensionality. This is important to understand for anyone who has dismissed machine-learning techniques as an exercise in overfitting. The architecture required to understand complexity has to be rich enough to contemplate all models of the world, and adaptive enough to live in the subset which is actually relevant at any point in time.

Predicting outcomes to events with no historical precedent is nearly impossible, which is why it's so important for investors to embrace a methodology that respects complexity.

My thanks to the Santa Fe Institute for including me in such good company and for creating a forum in which these important topics can be addressed. It's my hope that this historic time we're all navigating will remind us that the world we are living in is more complex than any of us could have imagined even a few months ago. The good news is that we are starting to better understand and develop the tools we need to harness the beauty of complexity. To quote Percy Shelley from two hundred years ago, "Poets are mirrors of the gigantic shadows the future casts upon the present." The scientific leaders at the Santa Fe Institute, past and present, are the poets of our millennium. As such, we hope they will reflect complexity's gigantic light upon all of us today and for years to come.

B. MILLER One of the most interesting statistics is that, if you go back forty or fifty years, and you look at the returns to investing in the top decile of free-cash-flow yield, that alone would give you 800 basis points per year of excess return over the S&P 500. It dwarfs any other factor or set of factors. And if you combine that with the top decile of share buybacks—companies with very high free-cash-flow yields—typically stock price would be therefore depressed, and they're buying back stock, so they're all getting capital in what appears to be the essential way, it jumps to 1200 basis points.

P. J. DAVIES We'll take another question over there.

AUDIENCE MEMBER Thank you. So, markets are complex adaptive systems and, in terms of the complexity, they're not that unusual; there are lots of complex systems out there. But, in terms of their adaptiveness, for the reasons that you mentioned, Dario, they're almost uniquely adaptive in that the feedback into the market is so much more rapid. Do you have thoughts on the application, not so much for markets as complex systems, but markets as adaptive systems?

B. MILLER For markets, I think there's a combination of positive and negative feedback. Dario mentioned reflexivity, which would be positive, but negative feedback's basically arbitrage, right? If two assets get out of line, there are smart people who are out there buying cheap ones; selling them dear won't bring them back in line. So, that mechanism is a very important one whether it's adaptive or not, but that certainly corrects those mispricings. But I think periodically we get reflexivity, where the system moves out of equilibrium and people are just all doing the same thing. That's also a phenomenon.

Whether you're adopting a new decision or rule, you're adapting your decision to reflect what's going on in the market. It could be someone who would be a natural seller just sits on their hands and does nothing; they don't sell, so they allow it to run or whatever it is. So, yeah, there's some sort of dynamics and there's adaptivity between positive and negative feedback, I think.

D. VILLANI There are a lot of applications for collective systems. That's why I was saying finance is the most . . . it's like, the condensed-matter theory of the field. It's where a lot of the new discoveries are going to happen. I really believe that. Somebody at MIT contacted me. One of the things our systems learns endogenously is the time scale at which to repay and let go of information.

Everybody has looked at economics. Even if I ask, "What is the beta of a stock?" You can compute better doing the regression over three months, six months, five years, ten years . . . which one would you use? At that time scale—people do, like, a moving window—it's key to be adaptive. And it changes all the time. There are fast learners, there are slow learners, and during periods of uncertainty, you shrink that time scale. Those error bars become big, and you become much more tentative. It's the interplay of being fast and slow. You know, we have a movie on that. The secret of great trading: you need to be fast in the sense of being adaptive and agile, but you can't be erratic. You can't get weak knees at every little spike move swaying you one way or the other. You need to be resilient through noise, but not obstinate. So, it's that line of fast and slow, that adaptiveness, that is really key. The key to all of that is to learn the time scale at which to retain or let go of information. It was the Department of Neural Science who wanted to do research with us on assessing the time scale as key for the adaptability in a living organism. That's one of the applications.

P. J. DAVIES Okay. I might draw it to a close there. Thank you very much for listening and engaging.

[APPLAUSE]

DAVID C. KRAKAUER Thank you very much. Thank you to the panel—thank you to all of you. It's been a long couple of days, very dense, lots of ideas, my brain is a bit of a snow globe. Let's just allow these ideas to settle, to see what's really going on. But I just want to thank everyone who spoke, and everyone who contributed. We're really grateful, particularly grateful to all the people who actually put this together.

AFTERWORD: THE FUTURE OF COMPLEXITY ECONOMICS: BETTER SOLUTIONS TO THE WORLD'S PROBLEMS

J. Doyne Farmer

The future of economics and the future of complexity economics are one and the same. The economics of the future will be inexorably driven toward complexity economics.

This transition needs to happen as fast as possible in order to create tools that can help guide us through the crises that the world faces. As I write, we are in the grips of the most economically devastating pandemic since the Black Death in the Middle Ages. For example, in Britain, where I live, economic activity decreased by an astounding 23.4 percent in the second quarter of 2020.[1] Unemployment in the US soared to almost 15 percent in April and remains at the high level of about 8 percent. The depth of the long-term damage to the economy that will be caused by the pandemic remains to be seen, but there is no question that it is significant, and it seems likely that the after-effects will linger for some time.

In these times, I shouldn't need to convince anyone how important economics is to our well-being. With better economic models, we could do a better job of guiding ourselves through crises like this one.

1 One should not draw conclusions from a single example. However, I cannot resist mentioning that our complexity economics model, which incorporated realistic features such as inventory management that other models neglected, predicted that the drop in UK economic activity during the second quarter of 2020 would be 22.5 percent. This was very close to the correct answer of 23.4 percent and much closer to the truth than the other forecasts.

OPPOSITE *Economic Curiosity: Victor Lustig*

The COVID-19 pandemic is just one of several big problems facing the world, including climate change, inequality, and financial crises.

Like an immense tsunami that is rolling toward us, climate change looms on the horizon and presents a far bigger problem than COVID. While climate change manifests itself as a physical phenomenon, it is important to bear in mind that it is caused by the economy. To mitigate climate change, we urgently need to rewire our economy to reduce carbon emissions to net zero as soon as possible. This means completely changing our energy infrastructure, which will in turn require major shifts in the complex supply chains that support it. These changes will reach deeply into the structure of the economy itself.

Inequality is ripping apart the fabric of society in many countries, particularly in the US and the UK. Although the problem has been building for the last fifty years, COVID is unfortunately making it much worse. The political divide that currently cripples the US government is caused at least in part by this widening chasm. We need to fix this as quickly as we can.

The aftermath of COVID threatens to cause yet another financial crisis. The damage to the real economy has put stress on businesses and households that is likely to be transmitted to banks and other financial institutions. This could reverberate back to the real economy, causing a feedback loop with the potential to deepen and prolong the damage for a long time to come.

Mainstream economics has not provided adequate guidance to solve these problems. Even worse, there are good arguments that it has contributed to creating them. A good example is neoliberalism, which has provided rationales for lowering taxes on the rich, dissolving labor unions, and other policies that have fostered the development of rampant inequality. Despite the fact that it is based on idealized arguments with little empirical support, its adherents have won numerous Nobel Prizes. Even if most economists are not neoliberals—indeed, most modern economists argue against it—the mere existence of these prizes calls into question the judgment of the profession.

Climate change provides perhaps the most glaring evidence that mainstream economics has failed us. In 2018, Bill Nordhaus received the Nobel Prize in economics for his work arguing that allowing the Earth to warm by 3.2°C provides an optimal trade-off between economic harm and climate change. In contrast, most physical scientists think this much warming could be disastrous. The problems with Nordhaus's analysis come from both an overestimate of the cost of combating climate change and an underestimate of the damage that it will cause, and the use of an oversimplified model based on unrealistic assumptions.

Our analysis indicates that we can convert most of the energy system within the next twenty years, save money as we do so, and keep global warming under 2°C.

Complexity economics provides a different perspective. Rupert Way, Penny Mealy, and I have analyzed the costs of the green energy transition, and we get the opposite answer to Nordhaus.[2] Our work builds on earlier work by Francois Lafond and some of my other collaborators at SFI and elsewhere[3] on forecasting technologies. Energy costs over the last century and a half—what one might call the fossil fuel era—have been surprisingly constant. Our data-driven forecasts show that renewables will very likely drive energy costs substantially lower than they have ever been. This means that converting the energy system to renewables is very likely a net savings; even if there were no climate change, we would be better off economically if we converted to renewables and did so quickly. Our analysis indicates that we can convert most of the energy system within the next twenty years, save money as we do so, and keep global warming under 2°C.

Why is our answer so different? This is at least in part because we follow the complex-systems maxim of building models from the

2 Way, Mealy, and Farmer (forthcoming).

3 See "References," p. 333 of this volume.

bottom up. Unlike Nordhaus, we do not rely on an aggregate model that depends on unrealistic assumptions such as rationality and utility maximization.

Financial crises provide another good example where insights from complexity economics have already been useful. The crisis of 2008 was caused by systemic risk. Systemic risk occurs when the decisions of individuals, which might be prudent when considered in isolation, combine to create risks at the level of the whole system that may be qualitatively different from the simple combination of their individual risks. By its very nature, systemic risk is an emergent phenomenon that comes about due to the nonlinear interaction of individual agents. To understand systemic risk, we need to understand the collective dynamics of the system that give rise to it. Complexity economics models have given us a deeper understanding of systemic risk, and are now starting to be deployed by central banks to help us monitor the financial system and mitigate future crises.[4]

If we are to get the necessary guidance to address big problems like climate change, inequality, and financial crises, we need models that can give us reliable answers.

Modern science and technology provide us with powerful tools to understand the world's problems and suggest effective solutions. We are now able to collect data on global economic activity at a remarkable level of detail. Computers have improved by a factor of roughly ten million since the middle of the twentieth century, and provide us with the power we need to process and understand this data. They potentially allow us to model the global economy from the bottom up with a remarkable level of fidelity and detail.

Mainstream economics has so far hardly scratched the surface of what could be done with all this data. The core concepts of utility

4 See Aymanns et al. (2018).

maximization and equilibrium on which mainstream economics is based date back to the nineteenth century. While these concepts are very useful for some purposes, such as designing auctions, they are not well suited to building models with the level of complexity required to take proper advantage of twenty-first-century technology in order to model the real world.

Complexity economics, in contrast, can guide us toward better solutions to the world's problems. This will be done by modeling the world from the bottom up. Complexity economists use a variety of tools, but, unlike mainstream economists, they do not shy away from using computers to simulate the economy (rather than proving theorems). Many individual agents are simulated, each with its own distinct behavior. Behavioral economics has made great progress in distinguishing real people from *Homo economicus* in the last few decades; complexity economics provides a natural vehicle to incorporate this knowledge into more realistic models. Many complexity economists (like me) are willing to abandon the sacred covenant of utility maximization in favor of better, empirically founded models of individual human behavior. In complexity economics models, economic phenomena emerge naturally from the interactions of individual agents.

Simulation and behavioral realism make a natural pair. Mainstream models based on utility maximization are solved via global optimization methods that typically can only be justified through "as-if" reasoning. In contrast, realistic behavioral models assume that agents process information locally and imperfectly, making these models easy to incorporate and run in simulations. Computer simulations make it possible to mimic a phenomenon with as much realism as needed, allowing models with *verisimilitude*. Good complexity economics models do not need to rely on "as-if" assumptions. While models are always idealizations, complexity economics promotes idealizations that do not force the omission of key features due to the constraints of mathematical tractability. "As-if" reasoning is replaced by "as-it-is" reasoning.

If we are to get the necessary guidance to address big problems like climate change, inequality, and financial crises, we need models that can give us reliable answers. This means the models need to incorporate all the features of the real economy that can affect these answers. Incorporating realism into models isn't easy. It will take some time before complexity economics models can achieve the level of realism that is needed. This is the only approach that can ever give us reliable answers to these big questions.

Complexity economics is presently on the fringe of economics, but it will eventually dominate it. Science succeeds only when its assumptions are realistic, i.e., when the resulting models have verisimilitude. The big problems the world faces demand realistic models, and the need for useful models favors complexity economics.

How will this happen? Because mainstream economics exerts a huge sociological pressure for conformity, it is difficult for complexity economics to take root in economics departments. Thus, I predict that complexity economics will emerge from other sources. These will include interdisciplinary work from within academia and by institutions such as central banks that have big incentives to get reliable answers. But I think the biggest force for change will ultimately be commercial. Complexity economics methods have the potential to give better forecasts and better analysis, and these are ultimately worth a great deal of money. This is already evident in the widespread and rapidly growing use of agent-based modeling in industry.

Science succeeds only when its assumptions are realistic, i.e., when the resulting models have verisimilitude. The big problems the world faces demand realistic models, and the need for useful models favors complexity economics.

Thus, I believe that the question is not *if* complexity economics will become a dominant paradigm in economics, but, rather, *when* this will happen. The forces driving this transition become increasingly larger as data collection improves and as computers become ever more powerful.

The gap between what is possible and what has been realized is already enormous, and it is widening. I think that within a decade we will start to see complexity economics models with important practical applications, whose performance is clearly superior to mainstream alternatives. And within twenty years they will start to enter the narrow confines of academia and become mainstream.

References

Rupert Way, Penny Mealy, and J. Doyne Farmer, "A Rapid Green Energy Transition is Cheaper than a Fossil Fuel Future." Under review; not public yet.

Farmer, J. D., and F. Lafond. 2016. "How Predictable Is Technological Progress?," Research Policy 45, 647–655.

Aymanns, Christoph, J. Doyne Farmer, Alissa M. Kleinnijenhuis, and Thom Wetzer. 2018. "Models of Financial Stability and Their Application in Stress Tests," in *Handbook of Computational Economics*, Volume 4, https://doi.org/10.1016/bs.hescom.2018.04.001.

APPENDIX

I cannot afford

The airport parking lot fee

SO NOW I LIVE HERE

—NICK PELTON, WEB DEVELOPER, SFI

ON SFI'S APPLIED COMPLEXITY NETWORK: A BRIEF HISTORY

William Tracy & Casey Cox

Organizations suffer when their leaders rely on simplistic or linear thinking to address complex problems. Yet most problems involving organizations, firms, governments, cities, pandemics, social media, markets, and environmental systems are inherently complex. Indeed, any system with networked adaptive agents can exhibit complex behavior. In this context, "complex" does not simply refer to a long or complicated list of independent but simple phenomena. Complex systems behave in ways that are qualitatively different from simple systems. Fortunately, complex systems exhibit universal mechanisms and behaviors which can be understood. The Santa Fe Institute is dedicated to exploring complexity's universal traits.

Originally established as The Business Network for Complex Systems Research in June, 1992, SFI's Applied Complexity Network (ACtioN) connects SFI researchers with a community of organizational leaders from foundations, governments, and private companies. These leaders spend the majority of their time confronting complex problems in domains as diverse as research and development, regulation, technology development, operations, organizational structure, IT, strategy, security, marketing, politics, international relations, and more. By exploring a variety of complex systems from different perspectives, researchers and ACtioN members gain new understandings and insights.

ACtioN member and SFI Trustee Toby Shannan offered some insight into this process of collaborative exploration in SFI's Spring 2017 *Parallax* newsletter. "I guess initially I kind of fell under the spell

OPPOSITE *Economic Curiosity: Gas Station Gourmet*

of complexity as a catchall for the problems we had at Shopify," said Shannan, Chief Operating Officer at Shopify. "As I read some of the scientific papers and gained a more technical understanding of the science, I began to make more tangible connections . . . It occurred to me that the interdisciplinary nature of complexity science extends to business. That's not true for most of science. The vocabulary of complex systems will be quite useful in the world of businesses as it deals with increasingly rapid change."

"This is now hands-on engagement and collaboration," SFI President David Krakauer said in the March/April 2016 *Update* newsletter. "We're learning from members and members are learning from us, and together, we're creating something new. ACtioN is about applying complexity insights to real problems. Our members are in a position to take these ideas the rest of the way."

A key goal of ACtioN is to identify the areas of complexity research that are ripe for application, or ready for input from practitioners. Towards this end, ACtioN hosts a variety of virtual and in-person topical meetings and an annual symposium (which this book reflects). These partnerships help complex-systems researchers better understand how complexity science can meaningfully impact the broader world.

In 2020, as COVID-19 drove interaction into the virtual space, the Applied Complexity Network began experimenting and adapting, expanding meetings to include new virtual forums for both presentation and collaborative discussion. A similar transformation has taken place in the educational opportunities made available to ACtioN members by SFI. This move online has significantly increased the accessibility of ACtioN events for members and researchers alike. As the Applied Complexity Network continues to evolve, we anticipate a new and exciting hybrid of virtual and in-person meetings.

SFI'S ANNUAL BUSINESS NETWORK/ APPLIED COMPLEXITY NETWORK SYMPOSIA 1994–2019

Nov. 11–12, 1994
Fall Business Network Symposium

Nov. 13–14, 1995
Annual Business Network Symposium

Oct. 21–22, 1996
Annual Business Network Symposium

Oct. 24–26, 1997
Business Network Annual Meeting

Oct. 16, 1998
An Exchange of Economic Ideas

Nov. 4–5, 1999
The Dynamics of Innovation

Nov. 3–4, 2000
Simplicity & Complexity:
A Global Perspective

Nov. 2–4, 2001
Global Economics

Nov. 1–2, 2002
Research–Santa Fe Style

Nov. 7–8, 2003
A Current Sampling of Santa Fe Institute–Style Research

Nov. 5–6, 2004
The Theory of Communication Networks: Flexibility, Robustness & Efficiency

Nov. 4–5, 2005
The Unreasonable Orderliness of Life
Feedback, Hierarchy & Emergence

Nov. 3–4, 2006
Infectivity

Nov. 2–3, 2007
Diversity Collapse: Causes, Connections & Consequences

Nov. 7–8, 2008
25th Anniversary: Open Questions in Science, Education & Business

Nov. 12–14, 2009
Multi-Dimensions of Evolution

Nov. 11–13, 2010
Complexity of Regulation

Nov. 3–5, 2011
Does the Individual Matter?

Nov. 1–3, 2012
Resilience

Oct. 31–Nov. 2, 2013
Big Data Meets Big Theory

Nov. 6–8, 2014
Complexity Economics

Nov. 12–14, 2015
Money & Currency:
Past, Present & Future

Nov. 3–5, 2016[1]
The Evolution & Complexity of Legal & Regulatory Systems

Nov. 3–4, 2017
The Complexity of Intelligence
New Science for Hybrid Intelligence

Nov. 9–10, 2018
The Emerging Frontiers of Invention

Nov. 8–9, 2019
New Complexity Economics

1 In 2016 the Business Network became the Applied Complexity Network (ACtioN).

SFI BUSINESS NETWORK/APPLIED COMPLEXITY NETWORK MEMBERS, PAST & PRESENT

Accenture PLC
Aflac
AIG
Air Force Research Laboratory (AFRL)
Alidade Incorporated
Anonymous Hedgefund
Anonymous Sovereign Wealth Fund
Argonne National Laboratory
Arience Capital
Baillie Gifford & Co.
Bank of the West
Barclays Capital
Barclays Global Investors
BlueMountain Capital Management
BMGI
Boeing Company
Booz Allen Hamilton
Boston Consulting Group
Bridger Capital
BT
Capital One
Centre for Liveable Cities
Cerelink
Change Science Institute
Cisco Systems
Clarivate Analytics
ClearBridge Investments
CNO Strategic Studies Group
Corteva Agriscience
Counterpoint Global
Credit Suisse
CSIRO
CustomerSat, Inc.
Davis Selected Advisers, LP
Deere & Company
Deloitte Center for the Edge
Deloitte Touche Tohmatsu
Descartes Labs
Dialog Group
Dynamo Capital, LLP
eBay, Inc.
Educational Testing Service
Eli Lilly & Company
Ethereum Foundation
Evidence Based Research (DOD)
Exxon Mobil Exploration Company
FedEx Corporation
Fidelity Investments
Fontainebleau Resorts
FX Palo Alto Laboratory
GE Healthcare
GKN Financial, LLC
Goldman Sachs
Google
Hakuhodo, Inc.
Harding Loevner
Hewlett Packard Enterprise Company
HH Capital Partners, LLC
HomeAway
Honda R&D Americas, Inc.
Honeybee Capital Foundation
HP, Inc.
HRL Laboratories, LLC
Humana Inc.
IBM
Icosystem Corporation
InnovationLabs, LLC
Insight Partners
Intel Corporation
Investor Analytics
IPSOS Science Center
Janus Henderson Investors
JMB Capital Partners
JWAC

Kinderhook Partners, LP
Lane Five Capital Management, LP
Lazard Asset Management
Legg Mason Capital Management
Lockheed Martin Company
Los Alamos National Bank
LPL Financial
Marathon Asset Management
Mars Incorporated
Matador Capital Management
McKinsey and Company
MedStar Institute for Innovation
Microsoft Corporation
Miller Value Partners
Morgan Stanley Investment Management
National Institute of Aerospace / NASA
National Semiconductor
Nativis, Inc.
NATS, LLC
Nokia
Northrop Grumman
NZS Capital
Office of Naval Research
OppenheimerFunds
Orange/Amena
Orange España/Amena
Orlando Health
Pfizer Inc.
Pioneer Hi-Bred International
Prediction Company
Principal Financial Group
Procter & Gamble
Psilos Group Managers, LLC
Putnam Investments
RAND Corporation
RCM
RedfishGroup
Repsol
RGM Advisors, LLC
SAC Capital Advisors, LLC
Sandia National Laboratories
SEAS DTC
SECOR Asset Management
Sherman Kent School for Policy Analysis
Shopify
Siemens
Sleep, Zakaria and Company, Ltd.
Soros Fund Management, LLC
Southeastern Asset Management, Inc.
State Farm Insurance Companies
Steelcase, Inc.
Sun Microsystems
Susquehanna International Group, LLP
Swiss Re Investors
Syngenta AG
Takeda Pharmaceuticals
Telecom Italia
The Harvest Group
The MITRE Corporation
The Omidyar Group
The Omidyar Network
The Thomson Corporation
Thinking Ahead Institute
Thomson Reuters
Thornburg Investment Management
Thornburg Mortgage Advisory
Thrivent Investments
Tortoise Capital Management, Inc.
Toyota Motor Corporation
Trust Company of the West
Tudor Investment Corporation
UBS
Undercurrent
Urban Mapping, Inc.
Veolia Environnement
Vizient, Inc.
Walt Disney Company
WCM Investment Management
We Mean Business
Willis Towers Watson
Ziff Brothers Investments

BUSINESS AS UNUSUAL

Janet Stites

reprinted from the SFI Bulletin, *Winter 2004*

A decade after its launch, SFI's Business Network has taken on a life of its own, acting as an agent to disseminate the theories and research of SFI researchers to the business community and, in turn, bringing back information to SFI. That is why a group of money managers meeting this fall in Newport, Rhode Island, have added the phrase "complex adaptive systems" to their financial dialogues; why a research director at one of the nation's premier funds-management firms is reading evolutionary theory at night after she puts her children to bed; and why SFI's External Faculty member David Stark's name and theory of "explore and exploit" are surfacing in a presentation given by a pharmaceutical executive.

Where did it start and where is it going? Network members are an elite group of self-selecting, open-minded business people from some of the world's largest and most forward-thinking companies. It has grown from an initial complement of five companies in 1992 to a current membership of more than 45 companies and government research groups, each of whom contributes $30,000 or more annually to support SFI's basic research agenda.

In return, Business Network members are invited to participate in SFI conferences and workshops, giving them the opportunity to network and interact with SFI scientists and look for ways to use SFI research at their own companies, while SFI benefits from the influx of new ideas.

This is a concept Michael Mauboussin considers daily. As Chief US Investment Strategist of Credit Suisse First Boston (CSFB), Mauboussin is charged with the task of absorbing and digesting data and information at a record pace, and then sculpting it into information the bank's investment team and clients can use—not for their own edification or to impress their friends—but to simply beat the market and make money.

CSFB joined the Business Network in 1997, and additionally, the company supports the research of J. Doyne Farmer, McKinsey Research Professor at SFI and founder of the Prediction Company. But Mauboussin, in his own work, is leveraging core concepts from SFI into his research, beginning with thinking of capital markets as complex systems. He began by studying W. Brian Arthur's theories on increasing returns, but continues to widen his scope, absorbing what he can on evolutionary biology and network theory and more.

After the 2003 East Coast blackout, Mauboussin put in a call to Columbia University and SFI External Faculty member Duncan Watts, who is an expert in network theory, to get his thoughts on the outage. "The blackout was essentially caused by a cascading failure in a large network," Mauboussin said. "I wanted to see what we could learn from that failure about networks and see how we could apply it to the financial markets."

Mauboussin is also known for his annual Thought Leader Forum, during which he draws on the work of SFI researchers. This year, the event, held in Newport, Rhode Island, featured Harvard-based geochemist and SFI External Faculty member Dan Schrag, who spoke about climate change. As well, Eric Bonabeau, former SFI postdoctoral fellow and founder and chief scientist of Icosystem Corporation, and Alpheus Bingham, a vice president with Eli Lilly and a member of the SFI Business Network, spoke at the Forum. In the past, SFI-affiliated speakers have included W. Brian Arthur, J. Doyne Farmer, John Holland, Duncan Watts, and Geoffrey West, among others.

Mauboussin remembers fondly the moment he learned about the Santa Fe Institute. "I was at an Orioles baseball game with Bill Miller (chief executive officer of Legg Mason Funds Management Inc.) in 1995," he says. "He told me I must be involved with SFI."

Bill Miller is a catalyst for many members to join the Business Network. Baltimore-based Legg Mason has been a member since the early 1990s under his leadership. A tireless advocate of the Institute and a vice chairman of the SFI Board of Trustees, Miller has embraced many of the theories imparted by SFI researchers. His own staff has followed suit.

Lisa Rapuano, director of research at Legg Mason Funds Management Inc., was somewhat confused when Miller first started sending her home evenings with books on evolutionary biology and network theory. But then she started attending the Business Network meetings at the Institute, and the disparity between finance and science began to wane.

When asked, Rapuano is initially hard pressed to come up with concrete anecdotes in which the firm has used research and information garnered from their time at the Institute. "You can't think of it in linear terms," she says. "What we have learned is to look at the market as an adaptive mechanism. We need to look for tools than aren't conventional. We need to develop a pool of alternative mental models to think about the market, companies, and economies."

Rapuano explains their strategy this way: "When we take our people out to SFI for the first time, they usually say, 'OK. That was interesting, but what am I supposed to do with the ideas on Monday?' We tell them, 'Nothing.' We tell them to absorb the ideas and let them enlighten their thinking."

For one example of how participating at SFI has enlightened Legg Mason's research theory, Rapuano points to the concept of "random search." "If you think about the way ants behave, they have a set of simple rules to go out and look for food," Rapuano says. "But one ant will run off to the side and look into a different spot, adding to the colony's overall robustness. When we do our own research, we keep in mind that we need to look everywhere. It might be as simple as a situation where you've typed in the wrong ticker, and instead of moving on, you stop and take a look at that company."

Similarly, Rapuano says the firm has incorporated the theory of "weak links" into their philosophy. "Research has shown that people get jobs through social networks—not usually through their friends, but through their friend's friends," she says."So, this is called a 'weak link.'"

"We try to look for what kind of connections make things happen,"she says."We go to conferences that aren't investment conferences." Making an even bigger commitment based on the "weak links" theory and network theory in general, the firm decided to sublet some space to a Baltimore hedge fund, betting they might garner something valuable from the liaison.

Perhaps the most important idea the researchers at Legg Mason have embraced is one of the most simple, yet fundamental to the work at the Institute: "We believe the market is a complex adaptive system with zillions of agents, with selfish objectives and excess," Rapuano says. "Return is difficult. You have to have a constant philosophy, but you must have an adaptive strategy. We need to be adaptive."

Rapuano, who tries to attend the Business Network meetings every year, says that she now looks for information not just from the SFI researchers, but also from her Business Network peers."There are really smart people at the meetings," she says. "You might sit next to the guy from Lilly and learn something you didn't know about pharmaceuticals."

Indeed, if that person is Alpheus Bingham, then you are most certainly likely to learn a great deal about pharmaceuticals. Bingham, a vice president of Eli Lilly and Company, has a way of putting a face on the otherwise intractable industry. In turn, Bingham has sifted through the myriad of information he gathers from the Institute and incorporated it into his work at Lilly.

Over the five years Bingham has been active in the Business Network, he believes it has helped him to reshape the structure in which corporate problems and challenges are framed. "It's allowed us to see alternatives that may have been less visible if stuck in traditional viewpoints," he says.

On the practical side, through connections made at SFI, Eli Lilly has incorporated agent-based modeling into its R&D processes, partnering with Eric Bonabeau's Icosystem to build modeling software, which helps the company track the progress of its research, and better understand its revenue flow. On the theoretical side, Bingham has been influenced by the theories of SFI-affiliated scientists such as Stuart Kauffman and David Stark.

A scientist himself, with a PhD in organic chemistry from Stanford University, Bingham is an ideal executive to be involved with the Business Network. He has long been incorporating technology into the research process; he is a visiting scholar at the National Center for Supercomputing Application at the University of Illinois, and former Chairman of the Board of Editors of *Research Technology Management Journal.* He believes the challenge for SFI's Business Network members is to find broader applications of complexity principles, to tap into the potential of the science. "Companies need to develop applications beyond simply using agent-based modeling programs," he says.

Perhaps no one, or no one company, has garnered as much from its affiliation with the Institute as Roger Burkhart, technical consultant to Deere & Company, who has graced the halls of the Institute for more than a decade.

Deere & Company joined the network in 1992. "We had begun developing the use of genetic algorithms for assembly-line scheduling and had developed an interest in adaptive techniques for both manufacturing and investment trading," Burkhart explains.

Two years later, Deere & Company "lent" Burkhart to SFI to participate for more than half a year on Chris Langton's Swarm Simulation team, which was developing the now well-known agent-based modeling and simulation platform Swarm, for modeling interactions of adaptive agents. Burkhart continues to help administrate the independent non-profit Swarm Development Group (www.swarm.org).

Having become something of a computer-based modeling evangelist, Burkhart's own projects explore the use of shared computer models across people and organizations, in such areas as product design and agricultural production. "One example," Burkhart explains, "would be integration of geospatial data from machines with agricultural production records to develop crop plans for a farmer. Many partners help collaborate with the farmer to develop and execute the plans, from input suppliers—seed, fertilizer, and chemicals—to agronomic consultants to output marketing channels."

Of late, the Business Network has become increasingly more reciprocal in nature. Many of the members gathered at the Institute last June for a topical meeting at which the Network members, not the scientists, had the microphone, addressing how they are applying research and information from SFI to their businesses.

Speakers represented a variety of industries—pharmaceutical, aviation, manufacturing, automotive, national laboratories, and, of course, high-tech—but were united in one goal of learning to harness the tools of complex-adaptive-systems research to help them with their own businesses.

Presenters included Bingham and Burkhart, as well as representatives from Intel, Sandia National Laboratory, Argonne National Laboratory, The MITRE Corporation, and Alidade Inc., among others.

Recently appointed SFI President Bob Eisenstein was impressed with the exchange of ideas. For the success of the program he credits the work of SFI staff members Suzanne Dulle and Susan Ballati. "Many problems studied at SFI are also problems of interest to the business community," he says. Eisenstein plans to make no major changes in the Business Network except to focus on bringing in more international firms. "We already have a significant foreign presence," he says. "But we want to connect to businesses in countries such as China and India as well."

One aspect of the Network Eisenstein wants to continue to emphasize is that the exchange is mutually beneficial. "We learn from the members just as they learn from us. Their input is valuable to us. In fact, sometimes there are problems they want to solve that turn out to be interesting problems for us. It's really a two-way street."

Visiting SFI Researcher José Lobo, who has been affiliated with the Institute since 1993, attended much of last spring's topical meeting, listened to the Business Network members' presentations, and participated in much of the dialogue. "There is a growing awareness among SFI researchers and the leadership of the Institute that the Business Network represents a great intellectual source that has remained largely untapped," says Lobo. He cites examples of intellectual exchange between Network members and SFI researchers. One notable one is in the area of biologically inspired software design, a project involving SFI Researcher Walter Fontana and physicist Ann Bouchard from Sandia National Laboratories.

Lobo also describes a new working group on Organizational Design, which was started by Bingham, Roger Burkhart, and SFI Researchers John Miller (also of Carnegie Mellon University), Jim Rutt, and Lobo. The group plans to host a session at the next Business Network meeting and has written a paper on the topic. "We hope that this group can evolve into a full-fledged research project at SFI," Lobo says.

Ultimately, SFI and its Business Network are very young organizations. Central to the Network's continued growth is an acceptance of complexity theory as a valid tool for business applications. In an odd way, the Internet boom and bust and ongoing sluggish economy has opened a door for new ideas and cutting-edge research like that coming out of the Institute.

CSFB's Mauboussin echoes this when reflecting on his tenure in the Network. "Since I first joined the Business Network, I have seen my peers open up to new ideas and begin to search for new formulas," he says. "The point is that we can't think about things in the same way anymore. I'm not saying SFI has the answers. I don't know. But I think there are potentially important ideas in the study of complexity."

Janet Stites is a freelance writer based in New York. She has written for OMNI Magazine, Newsweek, *and* The New York Times.

NEW COMPLEXITY ECONOMICS
2019 SYMPOSIUM AGENDA

FRIDAY, NOVEMBER 8, 2019

SFI Cowan Campus, Noyce Conference Room

9:00a **Agent-Based Modeling for Complexity Economics: Tutorial for ACtioN members & Trustees**
William Rand, North Carolina State Univ.; Anamaria Berea, Univ. of Central Florida & Blue Marble Space Institute of Science

Inn at Loretto

12:00p LUNCH AND REGISTRATION

1:00 **Introduction to New Complexity Economics**
David C. Krakauer, Santa Fe Institute

1:10 **How Complexity Economics Can Help The World**
J. Doyne Farmer, INET Oxford, University of Oxford & SFI

1:50 **Consumers vs. Citizens: Social Inequality and Democracy in a Big-Data World**
Allison Stanger, Harvard University, Middlebury College & SFI

2:30 BREAK

3:00 **Ergodicity Economics: A Trajectory**
Ole Peters, London Mathematical Laboratory & SFI

3:40 **Formal Markets and Informal Networks**
Matthew O. Jackson, Stanford University & SFI

4:20 **Communication and Coordination in Experiments**
C. Mónica Capra, Claremont Graduate University

5:00 COCKTAILS AND DINNER RECEPTION

7:00 KEYNOTE: **Complexity Economics: Why Does Economics Need This Different Approach?**
W. Brian Arthur, PARC & SFI

8:00 ADJOURN

SATURDAY, NOVEMBER 9, 2019

Inn at Loretto

8:00a BREAKFAST AND REGISTRATION

9:00 **Introduction**
David C. Krakauer, Santa Fe Institute

9:15 **Computation and Complex Economies**
Robert Axtell, George Mason & SFI; Joshua Epstein, New York University & SFI; Jessica Flack, Santa Fe Institute; Blake LeBaron, Brandeis University; John Miller, Carnegie Mellon & SFI; Melanie Mitchell, Portland State & SFI
Moderator: David C. Krakauer, Santa Fe Institute

10:30 BREAK

10:45 **Physics and Economic Systems**
J. Doyne Farmer, INET Oxford, University of Oxford & SFI; Ole Peters, London Mathematical Laboratory & SFI; Maria del Rio-Chanona, University of Oxford; Cosma Shalizi, Carnegie Mellon University & SFI; David Wolpert, Santa Fe Institute
Moderator: *David C. Krakauer, Santa Fe Institute*

11:45 LUNCH

1:00p **The Economic Organism**
C. Mónica Capra, Claremont Graduate University; Scott Page, University of Michigan & SFI; Rajiv Sethi, Columbia University & SFI; Geoffrey West, Santa Fe Institute
Moderator: *William Tracy, Santa Fe Institute*

2:00 BREAK

2:15 **Economic Architectures**
W. Brian Arthur, PARC & SFI; Eric D. Beinhocker, University of Oxford, INET Oxford & SFI; Matthew O. Jackson, Stanford University & SFI; Allison Stanger, Harvard, Middlebury College & SFI
Moderator: *William Tracy, Santa Fe Institute*

3:15 BREAK

3:30 **From Theory to Application**
Katherine Collins, Putnam Investments & SFI; Michael Mauboussin, BlueMountain Capital & SFI; Bill Miller, Miller Value Partners & SFI; Dario Villani, Duality Group
Moderator: *Paul J. Davies,* The Wall Street Journal

5:00 ADJOURN

SFI Miller Campus

5:30 COCKTAIL RECEPTION

ECONOMIC CURIOSITIES

The artwork in this volume is drawn from the table centerpieces at the symposium: mosaic images which documented economic novelties and endeavored to answer the question What is value?

Ambergris

Ambergris is a solid, waxy, flammable substance produced from a secretion of the bile duct in the intestines of sperm whales. It takes years to produce, and only an estimated one percent of sperm whales produce it. Freshly produced ambergris has a marine, fecal odor, but acquires a sweet, earthy scent as it ages.

It has been very highly valued by perfumers as a fixative that allows the scent to last much longer, and has historically been used in food and drink. A serving of eggs and ambergris was reportedly King Charles II of England's favorite dish. It has been used as a flavoring agent in Turkish coffee and in hot chocolate in eighteenth-century Europe.

Civet Coffee

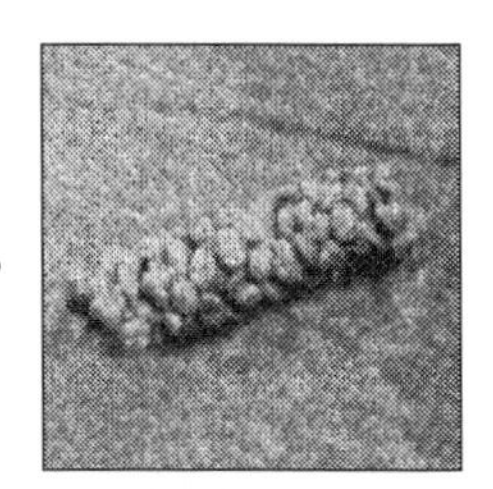

Kopi luwak, or civet coffee, is coffee that includes partially digested coffee cherries, eaten and defecated by the Asian palm civet (*Paradoxurus hermaphroditus*). Fermentation occurs as the cherries pass through a civet's intestines, and after being defecated with other fecal matter, they are collected. The traditional method of collecting feces from wild civets has given way to intensive farming methods that have raised ethical concerns about the treatment of civets.

Although kopi luwak is a form of processing rather than a variety of coffee, it has been called one of the most expensive coffees in the world, with retail prices reaching $700 per kilogram.

Francis Crick's Letter to His Son

"Dear Michael, Jim Watson and I have probably made a most important discovery," begins a letter written on March 19, 1953, by Francis Crick to his twelve-year-old son. The letter, auctioned at Christie's on April 10, 2013, for a record price of $6 million, is an unusual combination of the historic and the quotidian.

Signed "Lots of love, Daddy," it spells out each syllable of

DNA's full chemical name, "des-oxy-ribose-nucleic-acid," with the admonition to "read it carefully." At the same time, it contains the first written description of DNA as a code and of the mechanism of DNA replication—preceding the pivotal scientific articles about DNA's structure and its genetic implications by more than a month.

Gas Station Gourmet

Congruent with increasing income inequality around the world, significant profits are earned at opposite ends of the food quality spectrum. Gas stations are junk food meccas—hot dogs, soda, potato chips, energy drinks, cookies, and candy to fuel up for another stretch on the highway. These highly processed pre-packaged foods are a leading contributor to the profit margins of the largest food and beverage manufacturers in the US.

On the other end of the spectrum, there is immense profit potential in gourmet dining. Certain chefs have succeeded not only in creating incredible culinary experiences, but in building gourmet food empires—high-end junk food.

Leonardo da Vinci's Salvator Mundi

The *Salvator Mundi*, dated to c. 1500, was long thought to be a copy of a lost original painting by Leonardo da Vinci. It was rediscovered, restored, and displayed at the National Gallery, London, in 2011–12.

It is one of fewer than 20 known works by Leonardo, and was the only one to remain in a private collection. It was sold at auction for $450.3 million on November 15, 2017, to Prince Badr bin Abdullah, setting a new record for most expensive painting ever sold at public auction.

Lunch with Warren Buffett

For Warren Buffett, one of the richest men on the planet, time is precious. Because he is perhaps the savviest investor in history, anybody keen on making money by making smart investments will be eager to spend some time with him

Buffet does his bit for charity by auctioning his lunch time once a year. The winning bidder and up to seven friends can dine at the Smith & Wollensky steakhouse in Manhattan with Buffett, who says he will discuss anything apart from his next investments. This year, an anonymous bidder jumped on the opportunity and spent $4,567,888 to have lunch with Buffet. Over 20 years of auctions, Buffett, 88, has raised about $34.2 million for the Glide Foundation, a charity in San Francisco.

The Man Who Sold His Face

Forehead advertising made headlines in January and February 2005 when a 20-year-old man, Andrew Fischer, auctioned his forehead for advertising space on eBay. The winning company, SnoreStop, bid $37,375 for Fischer to display the company's logo via temporary tattoo on his forehead for 30 days. After this auction, Fischer became known internationally as the "Forehead Guy."

Fischer, who has since founded a viral marketing company named NURV.com and runs a website called PopMalt.com, recently posted his memoir about lessons learned from chasing fame. "Maybe there's more to being successful than being known for gimmicks," said Fischer, who wants his experience to be a cautionary tale to young people today obsessed with followers and likes in this social media age.

Once Upon a Time in Shaolin

Once Upon a Time in Shaolin is the seventh studio album by American hip hop group Wu-Tang Clan. It was limited to a single copy sold in 2015, and is the most expensive individual album ever sold, at $2 million. The album was recorded in secret over six years. A single two-CD copy was pressed in 2014 and stored in a secured vault at the Royal Mansour Hotel in Marrakech, Morocco, then auctioned through auction house Paddle8 in 2015. The winning bidder was American businessman Martin Shkreli. In March 2018, following Shkreli's conviction for securities fraud the previous year, a federal court seized assets belonging to him, including *Once Upon a Time in Shaolin*.

Red Paper Clip

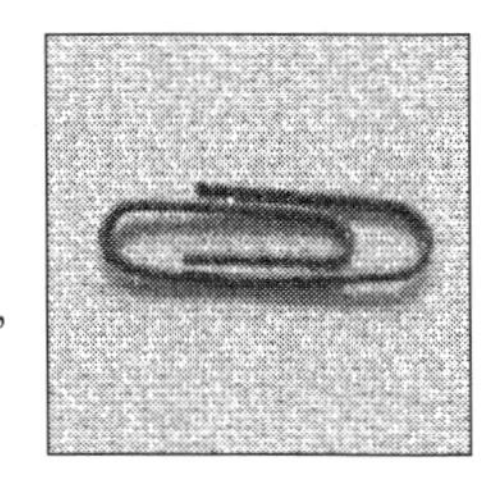

In 2005–2006, Canadian blogger Kyle MacDonald bartered his way from a single red paperclip to a house in a series of fourteen online trades over the course of a year. His trade began with an exchange of the paperclip for a fish-shaped pen, and went on to include a doorknob, a camp stove, a Honda generator, an "instant party," a Ski-Doo snowmobile, a two-person trip to Yahk, British Columbia, a box truck, a recording contract, a year's rent in Phoenix, Arizona, one afternoon with Alice Cooper, a KISS motorized snow globe, and a role in the film *Donna on Demand*. Finally, he traded the movie role for a two-story farmhouse in Kipling, Saskatchewan.

10,000 Bitcoins for a Couple of Pizzas

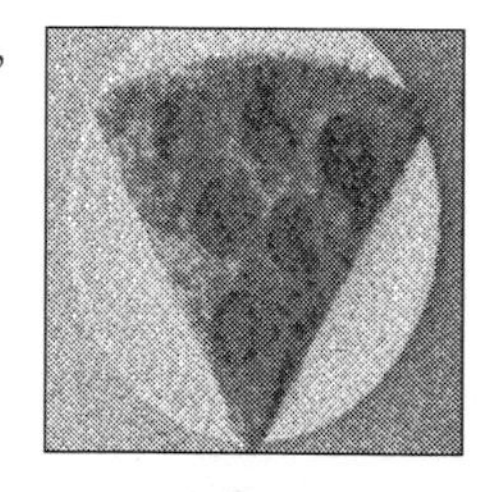

The world's first Bitcoin transaction, on May 22, 2010, exchanged 10,000 Bitcoins for two delivered Papa John's pizzas. The day, now known as Bitcoin Pizza Day, has become part of folklore, largely because of the price: the 10,000 Bitcoins, worth $41 at the time, reached parity with the US dollar just nine months later, making the two pizzas worth $10,000. Those Bitcoins would be worth $107,124,000 today. Despite the astronomical rise in the price of Bitcoin, it seems the Florida-based pizza recipient, Laszlo Hanyecz, is not fazed about his deal. "It wasn't like Bitcoins had any value back then, so the idea of trading them for a pizza was incredibly cool," Hanyecz told *The New York Times*.

Tulip Mania

During the tulip mania period of the Dutch Golden Age, prices for the new-to-Europe tulip bulbs reached extraordinarily high levels, having become a status symbol. In the 1630s, tulip prices spiraled upwards, prized by collectors beyond their obvious worth. One bulb of the most rare tulip variety, the *Semper augustus*, was advertised for 13,000 florins, the price of a mansion. But in 1637, the market for tulips in the Netherlands crashed nearly overnight. It is generally considered the first recorded speculative bubble.

Virgin Mary Grilled Cheese Sandwich

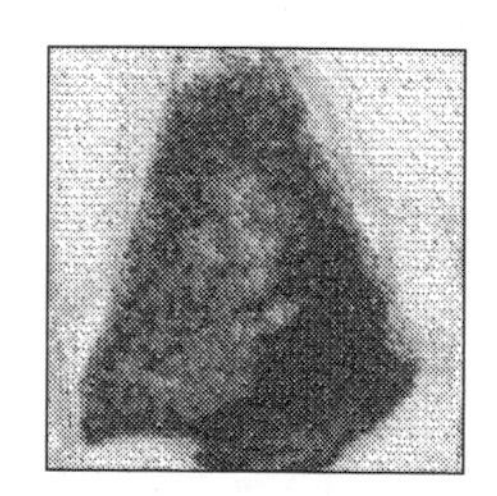

In 2014, a woman sold a 10-year-old grilled cheese sandwich on eBay for $28,000. The sandwich, which had a single bite taken from it, also bore the likeness of the Virgin Mary. Online casino GoldenPalace.com bought the sandwich from Diana Duyser. Duyser says she saw face staring back at her as she bit into the sandwich. She put the sandwich in a clear plastic box, kept it on her night stand, and reports that it never sprouted any mold.

Victor Lustig

Victor Lustig (1890–1947) was a highly skilled con artist from Austria-Hungary who undertook a criminal career that involved conducting scams across Europe and the United States. He is widely regarded as one of the most notorious con artists of his time. When the Great Depression hit, Lustig concocted a risky scam aimed at Al Capone, knowing that he faced certain death if he were to be found out.

SERIES EDITORS

WILLIAM TRACY is the Vice President for Applied Complexity at the Santa Fe Institute. Tracy oversees a suite of SFI programs that promote the application of complexity science beyond academia. He came to SFI from Rensselaer Polytechnic Institute, where he was the undergraduate program director for the Lally School of Management and a faculty member. He is proficient in Mandarin Chinese and formerly served as the Associate Director of SFI's CSSS–Beijing program. Tracy was at the World Bank before entering academia, where he focused on human developmental economics in Eastern Europe and Central Asia. He also has private sector and entrepreneurial experience in the US, China, and India. Tracy's personal academic work lies at the intersection of complex systems and strategic management, with a focus on how boundedly rational actors approach novel problems. He holds a PhD in management with a certificate in human complex systems from UCLA and a BA (cum laude) in economics from Swarthmore College.

CASEY COX is the Director of the Applied Complexity Network at the Santa Fe Institute. She spent her undergraduate years studying film at Syracuse University and the College of Santa Fe. After nearly fifteen years working in the film industry in Detroit and New Mexico, she pivoted her career, using her production skills to produce international science events and her networking skills to foster meaningful connections and collaborations between scientists and practitioners. Coming full-circle, Cox has just completed producing a pilot episode with a small team of scientists and filmmakers for a potential upcoming series on Complexity Science and the Santa Fe Institute.

VOLUME EDITORS

W. BRIAN ARTHUR is an External Professor at the Santa Fe Institute and Visiting Researcher in the Intelligent Systems Lab at PARC (formerly Xerox Parc) in Palo Alto. Arthur pioneered the modern study of positive feedbacks, or increasing returns, in the economy—in particular their role in magnifying small, random events in the economy and locking in dominant players. This work has gone on to become the basis of our understanding of the high-tech economy. In 2009 he published the book *The Nature of Technology: What It Is and How It Evolves*, an elegant and powerful theory of technology's origins and evolution.

Arthur is also one of the pioneers of the science of complexity. His association with the Santa Fe Institute goes back to 1987. He is a member of SFI's Founders Society, and in 1988 directed its first research program—work that has subsequently become the basis for Complexity Economics. He has served many years on SFI's Science Board and Board of Trustees. From 1983 to 1996, he was Morrison Professor of Economics and Population Studies at Stanford University, at the time of appointment, the youngest endowed professor at Stanford. Arthur is the recipient of the Schumpeter Prize in economics, the Lagrange Prize in complexity science, and two honorary doctorates. He earned his PhD from Berkeley in Operations Research and has other degrees in economics, electrical engineering, and mathematics

ERIC D. BEINHOCKER is a Professor of Public Policy Practice at the Blavatnik School of Government, University of Oxford, the Executive Director of the Institute for New Economic Thinking (INET) at the University's Oxford Martin School, and an External Professor at the Santa Fe Institute. INET Oxford is a research center devoted to applying leading-edge interdisciplinary approaches to issues including financial system stability, innovation and growth, economic inequality, and environmental sustainability. Beinhocker is also a Supernumerary Fellow in Economics at Oriel College.

Prior to joining Oxford, Beinhocker had an eighteen-year career at McKinsey & Company where he was a partner and held leadership roles in McKinsey's Strategy Practice, its Climate Change and Sustainability Practice, and the McKinsey Global Institute. Beinhocker writes frequently on economic, business, and public policy issues and his work has appeared in the *Financial Times, Bloomberg, The Times, The Guardian, The Atlantic, Newsweek, Democracy,* and he is the author of *The Origin of Wealth: The Radical Remaking of Economics and What It Means for Business and Society.* Beinhocker is a graduate of Dartmouth College and the MIT Sloan School, and is originally from Boston, Massachusetts.

ALLISON STANGER is the Technology and Human Values Senior Fellow at Harvard University's Edmund J. Safra Center for Ethics, the Russell Leng '60 Professor of International Politics and Economics at Middlebury College, New America Cybersecurity Fellow, and an External Professor at the Santa Fe Institute. She is the author of *Whistleblowers: Honesty in America from Washington to Trump* and *One Nation Under Contract: The Outsourcing of American Power and the Future of Foreign Policy*, both with Yale University Press. She is working on a new book tentatively titled *Consumers vs. Citizens: Social Inequality and Democracy in a Big-Data World.* Stanger's writing has appeared in *Foreign Affairs, Foreign Policy, Financial Times, International Herald Tribune, The New York Times, USA Today, US News and World Report,* and *The Washington Post*, and she has testified before the Commission on Wartime Contracting, the Senate Budget Committee, the Congressional Oversight Panel, the Senate HELP Committee, and the House Committee on Government Oversight and Reform. She is a member of the Council on Foreign Relations and received her PhD in Political Science from Harvard University.

CONTRIBUTORS

ROBERT AXTELL is Professor of Computational Social Science in the Department of Computational and Data Sciences at George Mason University and an External Professor at the Santa Fe Institute. He works at the intersection of economics and computer science. His most recent research attempts to recreate the US private sector from hundreds of millions of interacting agents who form firms, work described in his forthcoming book *Dynamics of Firms: Data, Theories, and Models* (MIT Press). He teaches courses on agent-based modeling, computational economics, and parallel computing. His research has been published in *Science*, *Proceedings of the National Academy of Sciences USA*, and leading field journals. He is the developer of Sugarscape, an early attempt to do social science with agent-based models, and co-author of *Growing Artificial Societies: Social Science from the Bottom Up* (with J. Epstein). He holds an interdisciplinary PhD from Carnegie Mellon University.

C. MÓNICA CAPRA is a Professor in the Department of Economic Sciences at Claremont Graduate University. Her areas of expertise are experimental economics, behavioral economics, and neuroeconomics. Capra is interested in decision processes and the role personality plays in shaping economic choices. Her contributions in behavioral game theory include the explicit modeling of introspection with error and the study of the effects of mood on decisions. Transdisciplinary studies constitute an important component of her work; she has collaborated with data scientists, neuroscientists, and psychologists. Capra is a graduate of Franklin & Marshall College and she earned her PhD in Economics from the University of Virginia.

KATHERINE COLLINS is the first Head of Sustainable Investing at Putnam Investments and Portfolio Manager for Putnam's Sustainable Leaders and Sustainable Future strategies, with approximately $6 billion in assets under management. She also serves as Vice Chair of the Santa Fe Institute Board of Trustees. Ms. Collins is the author of *The Nature of Investing* and *Month of Sundays* and founder of Honeybee Capital, an independent investment research firm focused on sustainable investment themes. Katherine serves on numerous boards, including Last Mile Health, Omega Institute, and Harvard Divinity School Dean's Council. She earned a Master of Theological Studies from Harvard Divinity School and a BA from Wellesley College, and is a CFA charter holder. Her closest neighbors in Massachusetts are thousands of honeybees.

PAUL J. DAVIES is a Senior Markets Reporter for *The Wall Street Journal* in London. He previously wrote about banking and finance for "Heard on the Street" in London. Before that he spent fourteen years at the *Financial Times* in a variety of finance-related roles. He is @PaulJDavies on Twitter.

JOSHUA M. EPSTEIN is a Professor of Epidemiology in the NYU School of Global Public Health and founding Director of the NYU Agent-Based Modeling Laboratory, with affiliated appointments at The Courant Institute of Mathematical Sciences, the Department of Politics, and is an External Faculty Fellow at the Santa Fe Institute. He is a recognized pioneer in agent-based modeling. Epstein's research applies mathematical and computational methods to complex social dynamics, including infectious and vector-borne diseases, urban disaster preparedness, contagious violence, the evolution of norms, economic dynamics, computational archaeology, and the emergence of social classes. His books include *Nonlinear Dynamics, Mathematical Biology, and Social Science*; *Generative Social Science*; *Agent_Zero*; and, with Robert Axtell, *Growing Artificial Societies*. He earned his BA from Amherst College and his PhD from The Massachusetts Institute of Technology.

J. DOYNE FARMER is Director of Complexity Economics at the Institute for New Economic Thinking at the Oxford Martin School, and is the Baillie Gifford Professor at Mathematical Institute at the University of Oxford, as well as an External Professor at the Santa Fe Institute. His current research is in economics, including financial stability, sustainability, technological change and economic simulation. He was a founder of Prediction Company, a quantitative automated trading firm that was sold to the United Bank of Switzerland in 2006. His past research spans complex systems, dynamical systems, time series analysis, and theoretical biology. He founded the Complex Systems Group at Los Alamos National Laboratory, and, while a graduate student in the '70s he built the first wearable digital computer, which was successfully used to predict the game of roulette.

JESSICA FLACK is a Professor at the Santa Fe Institute and director of SFI's Computation Group (C4). Research in C4 draws on evolutionary theory, cognitive neuroscience and behavior, statistical mechanics, information theory, dynamical systems, and theoretical computer science to study the roles of information processing and collective computation in the emergence of robust structure and function in biological and social systems. A central philosophical issue behind this work is how nature overcomes subjectivity inherent in information processing systems to produce collective, ordered states. Flack was previously founding director of University of Wisconsin–Madison's Center for Complexity and Collective Computation in the Wisconsin Institutes for Discovery. Flack's work has been covered by scientists and science journalists in many publications and media outlets, including the *BBC, NPR, Nature, Science, The Economist, New Scientist, Current Biology, The Atlantic,* and *Quanta Magazine.*

MATTHEW O. JACKSON is the William D. Eberle Professor of Economics at Stanford University, an External Professor at the Santa Fe Institute, where he is also on the Science Board, and a Senior Fellow of the Canadian Institute for Advanced Research. He was at Northwestern University and Caltech before joining Stanford, and received his BA from Princeton University in 1984 and PhD from Stanford University in 1988. Jackson's research interests include game theory, microeconomic theory, and the study of social and economic networks, on which he has published many articles and the books *The Human Network* and *Social and Economic Networks*. He also teaches an online course on networks and co-teaches two others on game theory.

Honors Jackson has received include the von Neumann Award, a Guggenheim Fellowship, the Social Choice and Welfare Prize, and the B. E. Press Arrow Prize for Senior Economists.

DAVID C. KRAKAUER is the President and William H. Miller Professor of Complex Systems at the Santa Fe Institute. His research explores the evolution of intelligence on Earth. He served as the founding Director of the Wisconsin Institutes for Discovery, the Co-Director of the Center for Complexity and Collective Computation, and Professor of Mathematical Genetics all at the University of Wisconsin, Madison. David has been a visiting fellow at the Genomics Frontiers Institute at the University of Pennsylvania, a Sage Fellow at the Sage Center for the Study of the Mind at the University of Santa Barbara, a long-term Fellow of the Institute for Advanced Study in Princeton, and visiting Professor of Evolution at Princeton University. Krakauer was included in *Wired Magazine*'s 2012 Smart List as one of fifty people "who will change the world," and *Entrepreneur Magazine*'s 2016 list of visionary leaders advancing global research and business.

BLAKE LEBARON is the Abram L. and Thelma Sachar Chair of International Economics at the International Business School, Brandeis University. LeBaron was at the University of Wisconsin 1988–1998, and also served as director of the Economics Program at the Santa Fe Institute in 1993. A Sloan Fellow and recent recipient of the Market Technicians Association Mike Epstein Award, he currently directs the Masters of Science in Business Analytics program at Brandeis, and is part of a Brandeis interdisciplinary research and teaching group interested in modeling dynamics in a wide range of fields. LeBaron has been influential both in the statistical detection of nonlinearities and in describing their qualitative behavior in many series. His current interests are in understanding the quantitative dynamics of interacting systems of adaptive agents and how these systems replicate observed real world phenomenon, as well as understanding some of the observed behavioral characteristics of traders in financial markets.

MICHAEL MAUBOUSSIN is Head of Consilient Research at Counterpoint Global, Morgan Stanley Investment Management, and Chairman of the Board of Trustees at the Santa Fe Institute. Previously held roles include Director of Research at BlueMountain Capital Management, Head of Global Financial Strategies at Credit Suisse, and Chief Investment Strategist at Legg Mason Capital Management. Mauboussin is the author of three books, including *The Success Equation: Untangling Skill and Luck in Business, Sports, and Investing*, and is co-author, with Alfred Rappaport, of *Expectations Investing: Reading Stock Prices for Better Returns*. Mauboussin has been an adjunct professor of finance at Columbia Business School since 1993 and is on the faculty of the Heilbrunn Center for Graham and Dodd Investing. He earned an AB from Georgetown University.

BILL MILLER is the Chairman and Chief Investment Officer of Miller Value Partners and Life Trustee and Chairman Emeritus of the Board of Trustees of the Santa Fe Institute. During his tenure as sole manager of the Legg Mason Value Trust, its performance exceeded S&P 500 benchmark index a record fifteen consecutive years.[1] He was named fund Manager of the Year in 1998 by Morningstar,[2] The Greatest Money Manager of the 1990s by *Money* magazine, Fund Manager of the Decade by Morningstar.com, was named by Barron's to its All-Century Investment Team and received the Sauren Golden Award in 2015 and 2017. Miller earned his economics degree from Washington and Lee University. Subsequent to graduation, he served as a military intelligence officer overseas and then pursued graduate studies in philosophy in the PhD program at The Johns Hopkins University. He received his CFA in 1986. A long-time supporter of the Santa Fe Institute, Miller established the Miller Omega Fund in 2016.

1 Legg Mason Value Trust-Class C beat the S&P 500 on an annual basis from 1990–2005. Bill Miller no longer manages the fund.
2 Morningstar's Award for "Domestic Equity Fund Manager of the Year 1988" recognizes portfolio managers who demonstrate excellent investment skill, the courage to differ from consensus, and the commitment to shareholders necessary to deliver outstanding long-term performance.

JOHN MILLER is a Professor at Carnegie Mellon, an External Professor at the Santa Fe Institute, and Chair of the Institute's Science Steering Committee. His research focuses on the complex adaptive behavior that emerges in key social systems drawn from anthropology, biology, economics, finance, and political science. He analyzes these systems using key ideas from the study of complex adaptive systems and methods ranging from human experiments to models based on mathematics and adaptive computation. The general approach of using computational models to explore social life is captured in his book with Scott Page on *Complex Adaptive Systems* (Princeton University Press, 2007) and in his book *A Crude Look at the Whole* (Basic Books, 2016) he describes a variety of applications of complex systems thinking to issues that arise in business, life, and society. He is currently exploring the origins of social behavior driven by simple models of evolving computations.

MELANIE MITCHELL is the Davis Professor of Complexity at the Santa Fe Institute. She is the author or editor of six books and numerous scholarly papers in the fields of artificial intelligence, cognitive science, and complex systems. Her book *Complexity: A Guided Tour* won the 2010 Phi Beta Kappa Science Book Award and was named by Amazon.com as one of the ten best science books of 2009. Her newest book is *Artificial Intelligence: A Guide for Thinking Humans.* Mitchell originated the Santa Fe Institute's Complexity Explorer project, which offers online courses and other educational resources related to the field of complex systems.

SCOTT PAGE is the John Seely Brown Distinguished University Professor of Complexity, Social Science, and Management at the University of Michigan, and an External Professor at the Santa Fe Institute. In 2011, he was elected to the American Academy of Arts and Sciences. His research focuses on the myriad roles that diversity plays in complex systems. Page received his PhD in Managerial Economics & Decision Sciences from Northwestern University. He has written five books, including The *Model Thinker*, *Complex Adaptive Social Systems* (with John Miller), and *Diversity and Complexity*, which explores the contributions of diversity within complex systems.

OLE PETERS is a Fellow at the London Mathematical Laboratory, the Principal Investigator of its ergodicity economics program, and an External Professor at the Santa Fe Institute. He works on different conceptualizations of randomness in the context of economics. His thesis is that the mathematical techniques adopted by economics in the seventeenth and eighteenth centuries are at the heart of many problems besetting the modern theory. Using a view of randomness developed largely in the twentieth century he has proposed an alternative solution to the discipline-defining problem of evaluating risky propositions. This provides a novel interpretation of expected utility theory and implies solutions to the 300-year-old St. Petersburg paradox, the leverage optimization problem, the equity premium puzzle, and the insurance puzzle. It leads to deep insights into the origin of cooperation, intertemporal discounting, and the dynamics of economic inequality. You can find him on Twitter under @ole_b_peters, and he maintains a popular blog, ergodicityeconomics.com.

MARIA DEL RIO-CHANONA is a Mathematics DPhil student supervised by Doyne Farmer at the University of Oxford. Before starting her PhD, Maria did her BSc in Physics at Universidad Nacional Autónoma de México 2011–2016, was a research intern at Imperial College London and Ryerson University, and worked as a data scientist for two consulting firms. Del Rio-Chanona's research interests are broad but focus primarily on developing a data-driven network model of the labor market to understand the impact of automation on employment. Additionally, she has worked on alternative methods to assess forecasts for renewable energy generation and has studied shock propagation in networks. As a research intern at the International Monetary Fund, she studied global financial contagion in multilayer networks. In general, Maria's research focuses on complexity economics, networks and shock propagation, agent-based models, and the future of work.

RAJIV SETHI is a Professor of Economics at Barnard College, Columbia University and an External Professor at the Santa Fe Institute. His research deals with information and beliefs. In collaboration with Brendan O'Flaherty, he has examined the manner in which stereotypes affect interactions among strangers, especially in relation to crime, policing, and the justice system. Their book, *Shadows of Doubt: Stereotypes, Crime, and the Pursuit of Justice*, was published by Harvard University Press in 2019. Rajiv is the 2020–21 Joy Foundation Fellow at the Radcliffe Institute for Advanced Study at Harvard, and has previously held visiting positions at Microsoft Research and the Institute for Advanced Study at Princeton. He has served on the editorial boards of several journals, including the *American Economic Review* and *Economics and Philosophy*.

COSMA SHALIZI is a Professor at Carnegie Mellon University, where he holds joint positions in the departments of Statistics and of Machine Learning, and co-directs the PhD program in Machine Learning, and an External Professor at the Santa Fe Institute. He studies methods of statistical inference for complex systems. After doing his PhD on the statistical mechanics of self-organization at the University of Wisconsin and SFI, he was a postdoc at the University of Michigan's Center for the Study of Complex Systems. Shalizi has worked extensively on complexity measures, self-organization, cellular automata, prediction and filtering of time series and of spatio-temporal data, heavy-tailed distributions, social network analysis, information flow in neural systems, causal inference from observational data, and collaborative problem solving. He still blogs sporadically at bactra.org/weblog/.

DARIO VILLANI is CEO and Co-Founder of Duality Group. Villani has managed multi-billion dollar portfolios within credit, interest rates, and commodities. Previously, he served as Global Head of Portfolio Strategy and Risk at Tudor Investment Corporation. He shared the 2016 Risk.net Buy-Side Quant of the Year Award, and has authored research papers in finance, theoretical physics, statistics and portfolio management. Villani holds a PhD in Theoretical Physics from Salerno University and a Master in Finance from Princeton University, where he also taught a popular course in trading and risk management.

GEOFFREY WEST is the Shannan Distinguished Professor and former President of the Santa Fe Institute and Associate Senior Fellow of University of Oxford's Green-Templeton College. He has a BA from Cambridge and a PhD from Stanford. West is a theoretical physicist whose primary interests have been in fundamental questions ranging from the elementary particles and their cosmological implications to universal scaling laws in biology and a quantitative science of cities, companies, and global sustainability. His work is motivated by the search for "simplicity underlying complexity." His research includes metabolism, growth, aging and lifespan, sleep, cancer and ecosystems, the dynamics of cities and companies, rates of growth and innovation, and the accelerating pace of life. West has lectured at many high profile events including TED and Davos. He has received many awards and his work has been featured in numerous publications, podcasts and TV productions worldwide. His work was selected in 2006 as a breakthrough idea by the Harvard Business Review and he was named to Time magazine's list of "100 Most Influential People in the World" in 2007. He is the author of the recent best-selling book *Scale*.

DAVID WOLPERT is a Professor at the Santa Fe Institute. The author of three books (and co-author of several more) and more than 200 papers, he is an IEEE fellow, has three patents, serves as associate editor at more than half a dozen journals, and has received numerous awards. He has more than 23,000 citations in a wide range of fields, including physics, machine learning, game theory, information theory, the thermodynamics of computation, and distributed optimization. He is a world expert on using nonequilibrium statistical physics to analyze the thermodynamics of computing systems; extending game theory to model humans operating in complex engineered systems; exploiting machine learning to improve optimization; and Monte Carlo methods. Prior roles include Ulam Scholar at the Center for Nonlinear Studies at Los Alamos National Laboratory, senior computer scientist at the NASA Ames Research Center, and consulting professor at Stanford University, where he formed the Collective Intelligence Group. His degrees in physics are from Princeton University and the University of California.

SFI PRESS BOARD OF ADVISORS

Nihat Ay
Professor, Max Planck Institute for Mathematics in the Sciences; SFI Resident Faculty

Sam Bowles
Professor, University of Siena; SFI Resident Faculty

Jennifer Dunne
SFI Resident Faculty; SFI Vice President for Science

Jessica Flack
SFI Resident Faculty

Mirta Galesic
SFI Resident Faculty

Chris Kempes
SFI Resident Faculty

Manfred Laublicher
President's Professor, Arizona State University; SFI External Faculty

Michael Mauboussin
Managing Director, Global Financial Strategies; SFI Trustee & Chairman of the Board

Cormac McCarthy
Author; SFI Trustee

Ian McKinnon
Founding Partner, Sandia Holdings LLC; SFI Trustee

John Miller
Professor, Carnegie Mellon University; SFI Science Steering Committee Chair

William H. Miller
Chairman & CEO, Miller Value Partners; SFI Trustee & Chairman Emeritus

Melanie Mitchell
SFI Davis Professor of Complexity, Science Board co-chair; SFI External Faculty

Cristopher Moore
SFI Resident Faculty

Sidney Redner
SFI Resident Faculty

Dan Rockmore
Professor, Dartmouth College; SFI External Faculty

Jim Rutt
JPR Ventures; SFI Trustee

Daniel Schrag
Professor, Harvard University; SFI External Faculty

Geoffrey West
SFI Resident Faculty; SFI Toby Shannan Professor of Complex Systems & Past President

David Wolpert
SFI Resident Faculty

EDITORIAL

David C. Krakauer
Publisher/Editor-in-Chief

Tim Taylor
Aide-de-Camp

Laura Egley Taylor
Manager, SFI Press

Sienna Latham
Editorial Coordinator

Katherine Mast
SFI Press Associate

THE SANTA FE INSTITUTE PRESS

The SFI Press endeavors to communicate the best of complexity science and to capture a sense of the diversity, range, breadth, excitement, and ambition of research at the Santa Fe Institute. To provide a distillation of discussions, debates, and meetings across a range of influential and nascent topics.

To change the way we think.

SEMINAR SERIES

New findings emerging from the Institute's ongoing working groups and research projects, for an audience of interdisciplinary scholars and practitioners.

ARCHIVE SERIES

Fresh editions of classic texts from the complexity canon, spanning the Institute's thirty years advancing the field.

COMPASS SERIES

Provoking, exploratory volumes aiming to build complexity literacy in the humanities, industry, and the curious public.

ALSO FROM SFI PRESS

InterPlanetary Transmissions: Stardust
David C. Krakauer & Caitlin L. McShea, eds.

Worlds Hidden in Plain Sight: The Evolving Idea of Complexity at the Santa Fe Institute, 1984–2019
David C. Krakauer, ed.

Emerging Syntheses in Science Proceedings of the Founding Workshops of the Santa Fe Institute
David Pines, ed.

For forthcoming titles, inquiries, or news about the Press, visit us at

WWW.SFIPRESS.ORG.

ABOUT THE SANTA FE INSTITUTE

The Santa Fe Institute is the world headquarters for complexity science, operated as an independent, nonprofit research and education center located in Santa Fe, New Mexico. Our researchers endeavor to understand and unify the underlying, shared patterns in complex physical, biological, social, cultural, technological, and even possible astrobiological worlds. Our global research network of scholars spans borders, departments, and disciplines, bringing together curious minds steeped in rigorous logical, mathematical, and computational reasoning. As we reveal the unseen mechanisms and processes that shape these evolving worlds, we seek to use this understanding to promote the well-being of humankind and of life on Earth.

COLOPHON

The body copy for this book was set in EB Garamond, a typeface designed by Georg Duffner after the Ebenolff-Berner type specimen of 1592. Headings are in Kurier, created by Janusz M. Nowacki, based on typefaces by the Polish typographer Małgorzata Budyta. For footnotes and captions, we have used CMU Bright, a sans serif variant of Computer Modern, created by Donald Knuth for use in TeX, the typesetting program he developed in 1978. Presenters' reflections are set in Embarcadero MVB Condensed, designed by Mark van Bronkhorst.

The SFI Press complexity glyphs used throughout this book were designed by Brian Crandall Williams.

SANTA FE INSTITUTE

COMPLEXITY GLYPHS

ZERO

ONE

TWO

THREE

FOUR

FIVE

SIX

SEVEN

EIGHT

NINE

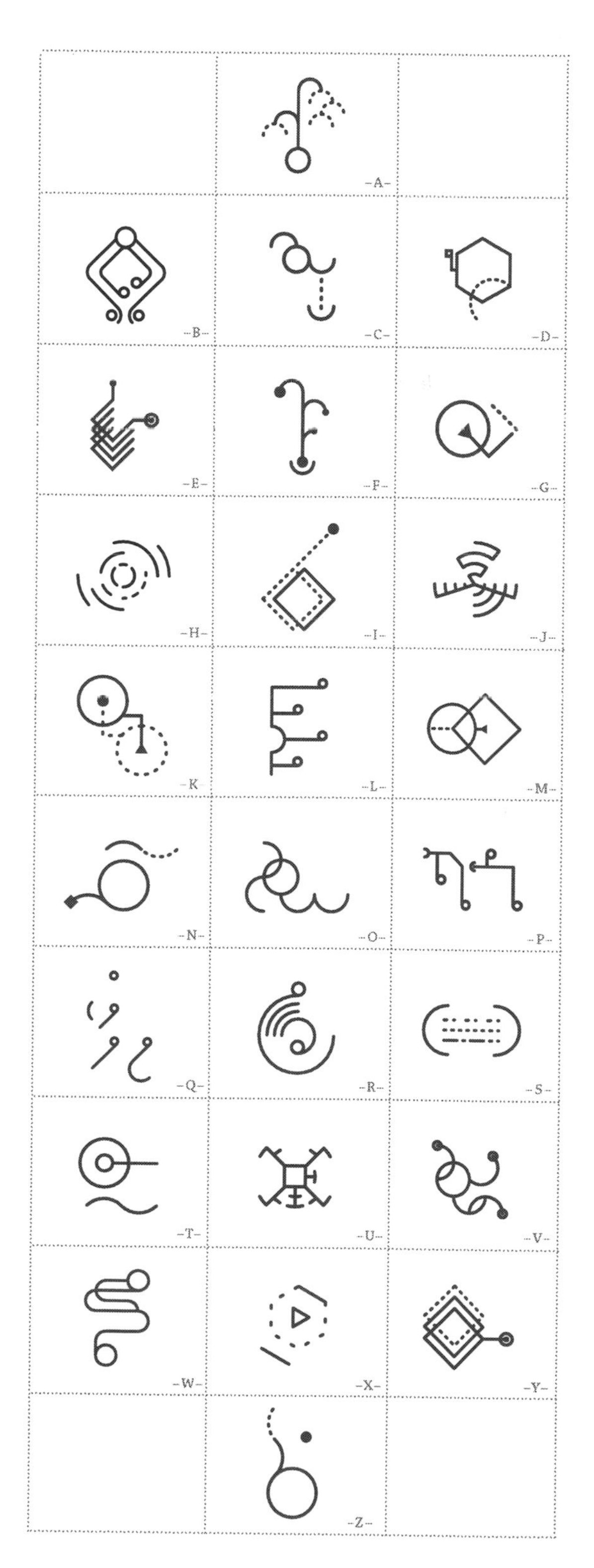

COMPASS SERIES